Software Forensics

Software Forensics

Collecting Evidence from the Scene of a Digital Crime

Robert M. Slade

McGraw-Hill

New York Chicago San Francisco Lisbon London Madrid
Mexico City Milan New Delhi San Juan Seoul
Singapore Sydney Toronto

The McGraw·Hill Companies

Library of Congress Cataloging-in-Publication Data on file

1 2 3 4 5 6 7 8 9 0 DOC/DOC 0 1 9 8 7 6 5 4

ISBN 0-07-142804-6

The sponsoring editor for this book was Judy Bass and the production supervisor was Pamela A. Pelton. It was set in Sabon by Patricia Wallenburg. The art director for the cover was Anthony Landi.

Printed and bound by RR Donnelley.

McGraw-Hill books are available at special quantity discounts to use as premiums and sales promotions, or for use in corporate training programs. For more information, please write to the Director of Special Sales, McGraw-Hill Professional, Two Penn Plaza, New York, NY 10121-2298. Or contact your local bookstore.

This book was printed on recycled, acid-free paper containing a minimum of 50% recycled, de-inked fiber.

To
Special, Treasure, Magic, and Shining

Contents

Introduction

Once Upon a Time ...

On Monday, August 18, 2003, the sixth variant of the Sobig virus family was found. Sobig was not technically sophisticated, as viruses go, and it was surprising that earlier variants had enjoyed the success that they did. Over the next few days, the Sobig.F variant (as it was called) surpassed all previous records in terms of the number of infected messages generated.

On Friday, August 22, researchers at F-Secure, a company in Finland that produces antiviral and other security software (and formerly known as Data Fellows), were analyzing the code of the virus. They were able to break the encryption of a section of code and find details of a payload which, up to that point, had been hidden.

The researchers determined that at 1900H UTC (noon, where I live) on Friday, computers infected with Sobig.F would try to connect to one of a number of servers, the IP addresses of which were coded into the virus. At that point, the virus would request a uniform resource locator (URL) from the server. The URL would direct the virus to go to a specific Web site, download an unknown program, and run it.

F-Secure researchers found the authentication code that the virus carried in order to prevent detection and downloading of the program. They connected to the servers, retrieved the URL, and tried to obtain the final program. It was not available at the URL specified. Presenting an incorrect URL would be an obvious way to prevent detection of the ultimate attack until the last minute.

The first point at which the researchers were sure enough to issue a warning about the situation was at about 1300H UTC, approximately six hours before the attack was due to start. Within that period, F-Secure, possibly with the assistance of other institutions, was able to contact the ISPs for these machines, and have them all shut down. One machine was left up as a honeypot, a form of trap for the people who intended to use it for the attack.

At this time, we do not know what the intention of the so-called "Stage 2" payload was, but the scheme shows evidence of very careful planning, and, given the extreme number of Sobig infections, it could have been very serious.

The design of this particular attack is interesting. The list of servers (there were 20) allowed for the possibility that a number of the servers might not be available at the time of the onslaught. The servers were spread across the United States, Canada, and South Korea, and had presumably been penetrated at some time previously, and were running at least some software under the control of the virus writer. The fact that the systems listed in the virus were only serving a URL, instead of a whole program, meant that the machines could answer requests from a very large number of infected machines, and allowed the person coordinating the assault to hide the actual program being downloaded until the last possible moment.

And Did Anyone Thank Them?

The foregoing is an example of the type of work in one area of software forensics. In this case, the task was to determine the intent and activity of a program. In other cases, we may wish to know what we can find out about the author of some code, or find out if theft of intellectual property has taken place because the software has been copied.

It is unlikely that you ever heard about the Sobig.F Stage 2 incident. It was a nonissue, and therefore not newsworthy, because the researchers found out about it in time and blocked the servers so that the rest of the attack could not take place. We will probably never know what the ultimate objective was, since the end program was not found. It may have been relatively innocuous, since the Sobig author seems to have been more interested in spamming and the resulting commercial opportunities rather than deliberate damage. (Email spam, of course, creates quite enough damage simply as a nuisance.)

But if the aim had been different, it could have dealt a very serious blow to the Internet, and everyone who depends on it.

Other Examples

While almost unknown outside of a small group of specialists, software forensics has had its successes.

The authors of the BRAIN virus, one of the earliest to hit IBM PCs in 1986 or 1987, were identified from the fact that they left their names, company name, and even address in the body of the code.

Analysis of only the code itself allowed experts to comprehend the functions of the 1988 Internet/Morris/UNIX Worm in only a few hours, and to understand which systems were vulnerable and how to fix them.

The author of the Den Zuk virus was identified by his ham radio license number in the program.

Ohio and Den Zuk, both written by the same person, were accurately identified as to the sequence in which they were written, as well as the intent of the programs, solely from examination of the software. These details were later confirmed in all respects by the author.

Even though the author of Suriv was never formally identified, examination of the software allowed virus researchers to determine that the Suriv family (which had three variants), Sunday, and the Friday the 13th viruses were all experiments working toward the successful Jerusalem-B virus.

Ongoing examination of the Empire and Monkey variants, as they appeared, allowed a researcher at the University of Alberta, aided by campus security, to slowly close in on the identity of the author.

Examination of the code of the eight relatively recent Klez variants indicates that the series is an ongoing diatribe against virus researchers and companies—by a virus author who felt that antivirus companies should have hired him as a consultant or programmer.

The author of the second variant of the Blaster (or Lovesan, or Poza) worm had it create a filename that was his online nickname, and also the name of a Web site that he had created and registered using his own real name and address.

The Background

I started out as a virus researcher.

In the case of viruses, of course, object code was usually all that was available. Even so, researchers were often able to determine a lot about a given piece of code. The first determination to make was whether a given program was malicious, and whether it was a virus. The next obvious question would be to try and determine who wrote the piece of malware. Sometimes virus writers made this an easy task, including names, addresses, and ham radio license call letters. However, it was also possible to find out whether one virus was modified from another, which came first, whether the modified version was created by the same author as the original, or whether someone else had used the original as a template. Researchers were often able to determine whether the programmer of a piece of software was a member of a specific linguistic, national, ethnic, cultural, or age group, and also the influence of various schools of programming on him or her.

(Sometimes the results were not as impressive. In the research over Den Zuk, there was a lot of debate as to whether this was a phrase from a foreign language. People were quite excited to find out that it might mean "sugar" or "the knife." Enthusiasm reached a fever pitch when we found out that, in one language, it could mean "the search," because the virus seemed to be intended to hunt down another specific virus. All of this was wrong: the author was a big fan of the "Danny Zuko" character from the movie *Grease*, and used Den Zuk as a nickname.)

For those of us in virus research, this attempt to find out more about virus writers by looking at the code of their products was fascinating, but completely academic. To the time of this writing, all the people convicted of writing viruses have admitted that they did it. Those virus researchers who made any money did so on the basis of creating or selling antiviral tools, or finding signatures for scanning for new viruses. There were no jobs in forensic programming, as we called it.

The Course

Some considerable time later, I was asked to put together a course to teach about viruses and virus protection. The phrase "forensic programming" caught the attention of the institution. So I was requested to write a syllabus for that topic. I was able to do so, but faced some challenges with respect to the materials to be presented to the students.

The Existing Books

The literature in print in regard to computer forensics concentrates on the recovery of data from seized computers. The emphasis is on slack or swap space, undeleting of erased files, and the chain of custody for presentation in court. More recently, there has been some interest in network forensics, and the ability to track messages and intrusions.

At the same time, some very significant and promising work is being done in regard to obtaining evidence from software. Unfortunately, the research is seen as of interest only to the academic community (the detection of plagiarism) or applications development (recovery of source code from object code). This material has yet to be formally introduced to the forensic community.

This Book

This work is meant to be that introduction. It is meant only as an introduction. Those who have been working with forensic programming, plagiarism detection, code analysis, source recovery, and even forensic linguistics will undoubtedly find the content simplistic. All I can say is that it was not meant for the experts.

Again, it is an introduction. Those who find the ideas presented here interesting need not think that they can read this work and then set up as forensic programmers. There are a number of skills required in the field that are quite beyond the scope of this book to address, such as skill in machine language programming and analysis. There are, of course, other books that can teach you about assembly language programming, but to deal with software forensics, you have to acquire a level of familiarity that only years of work will provide.

The fact that this book is just an overview of the topic means that the references and resources appendix will be significantly longer and more detailed than other parts of the book. This is to provide guidance to those who decide they want to pursue the topic further. At the same time, I am concerned that the book be of practical use to professionals, and therefore, I have avoided including academic papers that may not be generally available, in favor of books and Web sites that are. Those interested in the academic research on the topic could start by looking at CiteSeer (also known as ResearchIndex), http://citeseer.nj.nec.com/, which has a most amazing collection of research papers available online. Another starting point is the CERIAS library, at http://www.cerias.purdue.edu/coast/coast-library.html.

Please do not think that the book will be "hands-on" and "how-to." You will not find tables of signatures, nor will you find extensive sections of code. This book is intended as a conceptual and even abstract overview of the topic, and the related technologies that may be of use in software forensics. Given the nature of the material, tables of signatures would occupy whole volumes, and might also become dated very quickly.

The Audience (That's You)

This book is specifically intended for those involved in computer and digital forensics who need an introduction to the practices involved in, and information to be derived from, software forensics. It is also for security specialists who may want some background in regard to the analysis of malicious software. Those in the legal profession may need background to prepare cases that involve evidence from software.

This work is based on a course in software forensics, and could be used as a text for an introductory course in software forensics or forensic programming, or as an adjunct text in a course on computer or digital forensics. Software forensics as a practice requires a great deal of skill and experience. Therefore, this book is not a reference to software forensics. Such a tome would fill several volumes.

The Future

Software forensics is only just beginning, as a field, and is almost unknown outside of a few researchers and specialists. Hopefully this book will provide a base of information to a somewhat wider group of security practitioners and prompt more work in the discipline. I have undoubtedly missed areas that I should have covered and made errors in this text. For anyone who can point these out, I should be reachable as rslade@vcn.bc.ca or slade@victoria.tc.ca. Once I get this book finished, maybe I can update the course outline I've posted at http://victoria.tc.ca/techrev/fptoc.htm or http://sun.soci.niu.edu/~rslade/fptoc.htm. You point them out, I'll collect them, and maybe we can all do this again, sometime.

Acknowledgments

I would like to thank Carole McClendon, of Waterside Productions, for saving me all the time I didn't have to spend selling this book.

I would like to give a completely biased and subjective plug for Jeffrey Posluns and his very good folks at Security Sage (http://www.securitysage.com), who do virus and spam filtering, or sell you stuff to do it yourself. As I was trying to complete the final chapters of this book, I came under attack from someone in Atlanta with a Sobig infection and a broadband connection to the net. I was having to deal with the problems created by getting 6 to 10 megabytes of infected traffic per hour. Jeff graciously allowed me an account on his system, in order to use their filtering services. I forwarded my inundated account to rslade@computercrime.org (spiffy-cool address, huh?) and hey, presto—end of problem! Thank you, Jeff.

As always, this book would not have been completed without the backing of my organizer, cheering section, copy editor, writing critic, test subject, best friend, and wife, Gloria.

Software Forensics

1

Introduction to Software Forensics

You find an unknown file on a system for which you are responsible. You can tell that it is software, and an executable file, but you have no other information about it. Perhaps it is a vital, but undocumented part of your system. Or maybe it is a malicious program or utility, inserted into your system by an attacker. Having only the code itself, how can you tell?

A virus or other piece of malware is released anonymously and causes serious damage to systems around the world, including yours. Network traces provide no useful clues. Given only the malicious program itself, how much can you do toward finding the person who might be responsible for this potential disaster?

A controversy erupts over two software packages. The vendor of Program A claims that Brand X has been created using code that was extracted; in fact, stolen; from Program A, and that the interface and other unimportant details alone have been changed. Given only the two pieces of software, how certain can you be that plagiarism or theft of intellectual property has taken place? (For extra points, if one program was copied from the other, which came first?)

A problem is found in one of your modules in an enormous software project. Unfortunately, documentation as to exactly who wrote what has not been considered a priority, and is confused in any case by the fact that multiple coders may have worked on a given section. Your development department numbers in the hundreds. How do you find the programmer who wrote that particular section, in order to have the best insight into the problem and how to fix it?

If any of the foregoing scenarios led to litigation or prosecution, could you prove your contention in court?

These are examples of the types of questions that software forensics seeks to answer, by finding techniques for investigating and extracting evidence from software itself. To almost all users, most system managers, and possibly even a number of systems analysts, software, particularly in the object code state, is useful, but profoundly uninformative. It is, however, possible to determine a good deal of information from code, and this is the activity we will be examining.

This chapter is an introduction to the field of software forensics, its relationship to other areas of digital forensics, and an overview of the uses and practices of the discipline. Because it is an overview, sections of this chapter will reappear later in the book as full chapters, with expanded material and discussion in more depth. The content of this chapter, therefore, will be similar to that of the book as a whole, although the sequence will be different, to present a more logical and complete picture before we address specific details.

Digital Forensic Definitions

Computer forensics is commonly described as the study of evidence derived from computers. However, the literature on the topic concentrates primarily on the recovery and preservation for presentation as evidence, of data from computers that may have been used in the commission of some criminal activity. In fact, most books on the subject focus on the preservation aspects, fastening on documentation of the chain of evidence, and almost ignoring the technical aspects of the field.

The technical requirements are certainly rigorous. A computer forensic specialist should know a great deal about the normal functions of the operating system in question. For example, recent versions of the Microsoft Windows operating system may alter the data in as many as 400 files on the hard drive each time the computer is started up. This fact, and the identities and relevance of each of those files, must be taken into account if a computer has been restarted after it has been seized. That, though, is only a basic level of knowledge for computer forensics. The specialist must also be familiar with details of the internals of the operating system, such as common file formats for data storage. (UNIX and Windows systems will store identical plain text files with slightly different end-of-line indicators.) Professionals also need to understand the various file systems that may be encountered, in order to look for data in the slack space beyond the end of a file, blocks marked as unusable by the operating system, or the space in the zero track subsequent to the master and system boot records. True experts may need to deal with storage hardware, in order to access information hidden in the slack space between sectors, or in tracks outside of the normal range or spacing.

Once data have been recovered, computer forensics technicians may need to identify types of data structures, such as common archival or compression formats, from the internal structure of the file, and in the absence of normal indicators like the file extension. In some cases, cryptanalytic skills may be required to decipher simple encryption schemes, or try to recover keys for more sophisticated decryption.

All of the aforementioned skills may be relevant to software forensics because software, and particularly malicious software, may be hidden in a variety of locations, or may be encrypted with simple or subtle means.

The restriction of the computer forensics term to data recovery, and occasionally decryption, has been so complete that a new term has been coined to describe the more generic area of evidence from all forms of computer activity: digital forensics.

Network forensics is a major and growing field of digital forensics, and involves analysis of data from network logs and activity. A company can use network forensic analysis in order to detect intrusions or attacks launched against their network, generally from the Internet. This type of evidence can also be used to trace and track attackers or criminals who are using public systems such as the Internet.

With the increasing importance of communications to all aspects of computer work, network forensics will probably become just as important as the data recovery from computer forensics to the prosecution of cases. For example, it is no longer possible to rely on data from the local computer to determine whether a user has sent email because the email is just as likely to have been created on, and sent by, a remote computer system. No native indication of the message or its content may ever have been created on the local computer, even in the temporary work files that normally accompany such a process, since the processing did not happen on site. In particular, groups working together will likely be communicating over networks such as the Internet, and therefore, a combination of computer and network forensics will likely be used to piece together the whole story. Network forensics will probably be most useful in identifying groups involved in criminal or malicious activity.

Network forensics has only recently become a significant subject in security literature. In one sense, there is a body of material dealing with the topic: Discussions of firewall filtering lists, intrusion detection signatures, network attack indicators, and even spam prevention have presented a number of issues of interest to this field. Books dealing with network forensics on a broad scale, covering the entire area, and dealing specifically with the issues of identification, evidence, and presentation in court are only just beginning to appear.

The network forensic specialist requires knowledge of the internals of the network protocols and functions of the network under study. Again, not only the normal operations but, increasingly, abnormal sequences need to be identified. The task of the forensic expert may be somewhat eased by the near universality of the TCP/IP suite of network protocols, but is made more complicated by the growing number of network applications, and the particular uses and abuses that may be made of specially crafted packets.

Once again, network forensics is likely relevant to software forensics. Software may, and increasingly will, have or use communications functions. In addition, software, and particularly malicious software, may be created and distributed in such a way that copies never exist on the drive of the local computer. A number of recent network worms copied themselves solely to the memory of the computer. In cases such as these, copies of the software must be taken from the working memory of the computer, which presents a variety of problems. However, duplicates may also be obtained from logs or transcripts of packets sent or received over the network.

Software Forensics

Software, and particularly malicious software, has traditionally been seen in terms of a tool for the attacker. The only value that has been seen in the study of such software is in regard to protection against malicious code. However, experience in the virus research field, and more recent studies in detecting plagiarism, indicates that we can obtain evidence of intention, and cultural and individual identity, from examination of software itself.

Outside of the virus research community, forensic programming is a little known field. It involves the analysis of program code, generally object or machine language code, in order to make a determination of, or provide evidence for, the intent or authorship of a program. Software forensics, a relatively new addition to digital forensics, is the broader extension of this work. Software forensics involves the analysis of evidence from program code itself. Program code can be reviewed for evidence of activity, function, and intention, as well as evidence of the software's author.

Software forensics has a number of possible uses. In analyzing software suspected of being malicious, the information produced can be used to determine whether a problem is a result of carelessness, or was deliberately introduced as a payload. Information can be obtained about authorship and the culture behind a given programmer, and the sequence in which related programs were written. This can be used to provide evidence about a suspected author of a program, or to determine intellectual property issues. The techniques behind software forensics can sometimes also be used to recover source code that has been lost.

Software forensics generally deals with two different types of code. There is source code that is relatively legible to people. Analysis of source code is often referred to as code analysis, and is closely related to literary analysis. Analysis of object, or machine code, is generally referred to as forensic programming. We will examine the different types of code, and how software is produced, in Chapter 3, dealing with the objects of software analysis.

Literary analysis has contributed much to code analysis, and is an older and more mature field. It is referred to, variously, as authorship analysis, stylistics, stylometry, forensic linguistics, or forensic stylistics. More details on this type of work will be discussed in Chapter 9.

Objectives and Objects of Software Forensics

Historically, the virus research community has used forensic programming for a variety of purposes. First and foremost, of course, was to determine the intent of a program. Was this piece of code actually a virus, trojan, or other piece of malware, or had someone merely blamed it for some unrelated event? Various methods are used for this type of assessment, ranging from "black box" execution of the program to disassembly and decompilation. The tools, software and otherwise, used in this type of work will be described more fully in Chapter 3, along with the discussion of software code itself.

Commonly, in virus research, an attempt is made to determine whether a virus exists in other versions, or belongs to an existing family of viruses. Indications can be found in a direct analysis of the object code, analysis of a disassembly, or a review of text strings and messages that may be found in the code. With slight modifications of code, and the change only of text, specific triggers, or minor functions, the program might be considered a variant, and assigned a numeric or letter code, but identified with the original name. If structural changes have been made or new functions added, a virus may be assigned its own name, but noted to be part of a specific family.

In regard to families, an attempt is generally made to sequence the different variants. Even when a single author is involved, an analysis of the code can determine how the program developed. If we have a sequence of programs from a single author, we can potentially glean even more information that might help us identify the programmer.

Identity

This brings us to another objective for forensic programming and software forensics. From the earliest appearances of trojan horse programs on bulletin boards, there has been an interest in finding the authors of malicious software. In some cases, viruses have contained names, addresses, company names, email addresses, Web sites, and even ham radio license identifiers, either in plain text or in various forms of encryption. Other information can be obtained from indications in the code that signify that two programs were written by the same author, or that an author is a member of a group. As well, stylistic or stylometric analysis of messages and text may provide information and evidence that can be used for identification or confirmation of identity.

Individual Identification

I recently spoke at a conference where the section in which I was presenting was titled "Electronic Fingerprints." The term "electronic fingerprint" is particularly well chosen in regard to the identification of individuals as a result of this type of analysis. Physical fingerprint evidence frequently does not help us identify a perpetrator in terms of finding the person once we have a fingerprint. However, a fingerprint can confirm an identity, or place a person at the scene of a crime, once we have a suspect. In the same way, the evidence we gather from analyzing the text of a message, or a body of messages, may help to confirm that a given individual or suspect is the person who created the fraudulent postings. Both the content and the syntactical structure of text can provide evidence that relates to a particular individual.

Programmers have styles in the same way that writers have styles. Code may be sloppy or optimized, and, if optimized, may be optimized for either processor cycles or memory space. Programmers will have preferences for lookup tables or algorithmic methods, and for different types of loop structures. There will be other characteristics that programmers use, either consciously or unconsciously. Taken together, these can be used to compare a sample of program code to a body of such code that a programmer is known to have produced.

Sometimes, of course, our job is easier. Programmers, even of malicious code, have been known to "sign" their code in some way, with a nickname, their real name, addresses, email addresses, and Web sites.

Group Identification

Some of the evidence that we discover may not relate to individuals. Some information may relate to a group of people who work together, influence each other, or are influenced from a single outside source. These data can still be of use to us, in that they will provide us with clues in regard to a group that the author may be associated with, and may be helpful in building a profile of the writer in terms of the intersection of his membership in multiple groups.

Groups may also use common tools. One area we need to investigate in regard to program code involves programming environments that may generate or partially generate code for the programmer.

Compilers also have specific signatures, sometimes in a header area of the program, and sometimes in terms of the translation of source code into object, or optimization provided by the compiler program itself. Other types of tools, such as text editors or databases, may be commonly used by groups and provide similar evidence.

In analyzing software, one can find indications of languages, specific compilers, and other development tools. Compilers leave definite traces in programs, and can be specifically identified. Languages leave indications in the types of functions and structures supported. Other types of software development tools may contribute to the structural architecture of the program or the regularity and reuse of modules.

It is also possible to trace indications of cultures and styles in programming. A very broad example is the difference between program design in the Microsoft Windows environment and the UNIX environment. Windows programs tend to be large and monolithic, with the most complete set of functions possible built into the main program, large central program files, and calls to related application function libraries. UNIX programs tend to be individually small, with calls to a number of single-function utilities.

There are, of course, numerous examples of cultural influences in programming that are visible in user interfaces. It was fairly obvious to note the bias toward LISt Processing (LISP) programming that was evident in the Logo programming language, and the predisposition toward the University of California at San Diego (UCSD) P-system editor that clearly drove the developers of the Wordstar word processor (among others). The present day dominance of the Microsoft Windows interface has tended to homogenize interface choices. Yet even this can be seen as a cultural artifact: The inclination toward the Windows (originally the IBM Common User Access [CUA]) interface has become so commanding that developers go to extraordinary lengths to include "File," "Edit," "View," and "Tools" menus even on programs that have no need for those kinds of functions.

Programming and design cultures are clearly evident in malware. As a simplistic example, the early distributed denial of service (DDoS) tools could merely have opened a characteristic port that could be scanned for, but almost all were designed to "announce" availability once a machine had been compromised. The announcement could have been made through a variety of channels, and even anonymous ones, but Internet relay chat (IRC) was the one most commonly used. Again, DDoS client or agent programs (commonly called "zombies") could have been employed by having them "listen" for commands on IRC or Usenet newsgroups, but the authors all preferred to have the attack controller send attack commands directly to the agents.

In some cases, of course, similarity of code or design does not indicate influence as much as direct copying. Virus variants, for example, tend to be related merely because a virus "author" will simply take an existing virus and make minor variations to the code. (In many cases, the code is not changed at all: The new "programmer" will modify text strings, or will throw no operation [NOP] codes into the program—making no functional change.) Yet it is also possible to see where ideas and functions have been taken from one or more sources and added to a program. This is especially clear when the function is coded in a slightly different way.

Evidence of cultural influences exists right down to the machine code level. Those who work with assembler and machine code know that a given function can be coded in a variety of ways, and that there may be a number of algorithms to accomplish the same end. It is possible, for example, to note, for a given function, whether the programming was intended to accomplish the task in a minimum amount of memory space ("tight" code), a minimum number of machine cycles (high performance code), or a minimal effort on the part of the programmer (sloppy code).

More material and examples of cultural programming indicators will be covered in Chapter 7.

The blackhat community—hackers, crackers, phreaks, and other doodz. We will be dealing with the blackhat community in greater depth in Chapter 2, as well as malicious software in Chapter 6. (I use the term "blackhat" to avoid arguments about the "true" definition of a hacker.) However, it is instructive to look at the rough ideas we have been able to obtain about the groups of intruders and writers of malicious software. For this information, we are all indebted to researchers such as Sarah Gordon, Dorothy Denning, Ray Kaplan, and, more recently, the members of the Honeynet Project.

First, I should point out that the blackhat community is extremely fragmented. Not only are there different groups, often at odds with each other, but their types of activities differ. Some are trying to break into or intrude upon computer systems or networks. Others specialize in gaining unauthorized use of telephone switches and systems, frequently for the purpose of obtaining or even reselling phone service. Some are primarily interested in damaging or corrupting files, particularly in public ways, such as defacing Web sites. A great many of the blackhats in general, and probably the largest majority, really have very little idea of the technology that they are using, having obtained packaged programs or scripts that they operate without really understanding the functions, or appropriate uses. Blackhats who create programs of any type are actually relatively rare. A number do make slight modifications to the creations of others, usually functionally insignificant changes to viruses, which are widely available because of their reproductive function. There are, of course, those who are primarily interested in making illegal copies of commercial software. And, at every level, there are people who "wannabe" more respected in the blackhat community, but lack even minimal skills.

It may be important to examine the commonly presented justifications for blackhat activity for two reasons. First, the material does demonstrate something of the mindset and philosophy of the members of the community, and such a philosophy can sometimes be evident in programming style. Second, some of these justifications may be presented, quite seriously, as arguments against the activity of software forensics in general. However, as this is merely an introduction to the process of software forensics, I will leave analysis of motivation for another time.

General characteristics. Blackhats, and particularly writers of malware and viruses, tend to be young, and almost invariably male. Despite occasional speculations on the addictive nature of "hacking," they usually "grow out of" the virus writing game after a few years.

Virus and malware researchers tend to be dismissive of the technical abilities of virus writers. There exist virus writers who write competent code; there are many more who do not. The general public and the media, of course, continue to be fascinated by the image of the mythical "boy genius" running rings around the authorities. The blackhats like this cliché, too, and many go to some lengths to encourage the stereotype, whether or not they believe in it.

Malware writers don't understand or prefer not to think about the consequences of malicious software for other people, or they simply don't care. Recently, one researcher compared the characteristics of the blackhat community with those of people who fall somewhere in the range between an admittedly ill-defined "normal" and those suffering from full-blown autism. Autistic individuals tend to perceive and interpret the world in an idiosyncratic manner.

Malware authors draw a false distinction between creating malicious software and distributing it. They eschew any responsibility for the damage caused by their creations. In particular, they believe it is the responsibility of the victim to defend him or herself from encroaching malware, not the responsibility of the creators to keep their handiwork away from systems other than their own. Targets and victims of attacks are typically dehumanized in blackhat writings, described as losers who do not deserve to own a computer. There is also projection and displacement of guilt, frequently expressed in terms justifying security breaking activities because vendor X makes lousy software or large corporations are doing bad things.

Blackhat products. Most of the end result of blackhat activity consists of compromised systems, defaced Web pages, and pointlessly consumed bandwidth. Overall, this might be of interest to those investigating network forensics, but isn't of much use for us in software forensics. However, attack tools, DDoS kits, trojans, viruses, worms, remote access trojans (RATs), and other forms of malware are.

We will, of course, want to find out as much as possible about what the specific piece of malware does. We also want to know about the author, if we possibly can. Knowing about the broad classes of malicious software can be helpful in pointing us at the general functions to look for. Knowing the class of malware may also help us to identify the author because blackhats tend to be just as specialized as any other type of programmer.

Other Objects of Study

In the case of malware, those in software forensics are primarily concerned with finding out what the program does, and who wrote it. Frequently, this might be a concern with ordinary application software. Generally speaking, with regular software, we will know part of the answer, but need to uncover the rest.

One situation that may arise needing a forensic determination is in the case of intellectual property. It is possible that multiple claims of authorship may be made for a particular piece of software. It should be relatively easy to compare a specific program against two or more known bodies of work, and ascertain which of a number of authors has written the disputed application. In the case of multiple authorship of a single program or module, this would be more difficult, but forensic linguistics has, in some cases, been able to distinguish between multiple authors, even down to the level of an individual sentence. In any case, we should be able to conclude fairly easily whether multiple authors were involved. Plagiarism detection is already well established as a technology, and there are a number of automated tools that can help us in this regard.

The technologies employed in software forensics have uses in software development itself, and, indeed, some of them originated there. Reverse engineering has been a common practice for years in system development. Software forensics performs much the same function, albeit sometimes at different levels of detail. Disassembly and decompilation tools may be able to assist in application development, recovering, for example, source code for legacy systems where such has been lost over the years.

As noted, the tools used in software forensics are generally utilities employed in programming. At this point, it may be helpful to list the instruments that can be beneficial in forensic endeavors.

Software Forensic Tools

Before listing the tools themselves, some brief background on the programming process might be in order.

The Process

In the beginning, of course, programmers created object (or machine, or binary) files directly. The operating instructions (opcodes) for the computer, and any necessary arguments or data, were presented to the machine in the form that was needed to get it to process properly. Assembly language was produced to help with this process: There is a fairly direct correspondence between the assembly mnemonics and specific opcodes, at least the assembly files are formatted in a way that is relatively easy for humans to read, rather than as strings of hexadecimal or binary numbers.

With the advent of high, or at least higher, level languages, programming language systems split into two types. High-level languages are those where the source code is somewhat more comprehensible to people. Those who work with C or APL may dispute this assertion, of course. The much maligned COBOL is possibly the best example: The general structure of a COBOL program should be evident from the source code, even for those not trained in the language.

Compiled languages involve two separate processes before a program is ready for execution. The application must be programmed in the source (the text or human readable) code, and then the source must be compiled into object code that the computer can understand: the strings of opcodes. Those who actually do programming will know that I am radically simplifying a process that generally involves linkers and a number of other utilities, but the point is that the source code for languages like Fortran and Modula cannot be run directly: It must be compiled first. (It is, of course, dangerous to make such statements: Undoubtedly some completist computer language historian will be able to identify Fortran and Modula interpreters, of which I am totally unaware.)

Interpreted languages shorten the process. Once the program has been written, it can be run, with the help of the interpreter. The interpreter translates the source code into object code "on the fly," rendering it into a form that the computer can use. There is a cost in performance and speed for this convenience: Compiled programs are "native" or natural for the central processing unit (CPU) to use directly (with some mediation from the operating system), and so run considerably faster. In addition, compilers tend to perform some level of optimization on the programs, choosing the best set of functions for a given situation.

However, interpreted languages have an additional advantage: Because the language is translated on the machine where the program is being run, a given interpreted program can be run on a variety of different computers, as long as an interpreter for that language is available. Scripting languages, used on a variety of platforms, are a good example. JavaScript applets, for instance, may be embedded in Web pages, and then run in browsers that support the language regardless of the underlying computer architecture or operating system. (JavaScript is probably a bad example to use when talking about cross-platform operation, since a given JavaScript program may not even run on a new version of the same software company's browser, let alone one from another vendor for another platform. But it is supposed to work across platforms.)

As with most other technologies where two options are present, there are hybrid systems that attempt to provide the best of both worlds. Java, for example, "compiles" source code into a sort of pseudo-object code called byte-code. The byte-code is then-processed by the interpreter (called the Java virtual machine or JVM) for the CPU to run. Because the byte-code is already fairly close to object code, the interpretation process is much faster than for other interpreted languages. Because byte-code is still undergoing an interpretation, a given Java program will run on any machine that has a JVM.

(Java does have a provision for direct compilation into object code. So do a number of implementations for interpreted languages, such as BASIC. I think the language designers do this purely in order that explanations such as this cannot make straightforward statements.)

The Products

What we get for analysis depends, of course, on how the program was developed. If it was machine language programming, assembler, or a compiled language, we get an object code file for analysis. In the case of assembler or compilation, we may also have a copy of the assembler or high-level language source code. If we have an interpreted language used for development, we have a copy of the source code of the program. (For the purposes of software forensic analysis, partially compiled objects such as Java byte-code can be considered under the same type of analysis as object code. Also, source code, where available, can be assessed in a similar manner regardless of whether the language used was a compiler or an interpreter.)

However, the development system still has some ways to make our analytical task more difficult.

Complicating Factors

When a program is compiled or assembled, all comments (unless they are handled in special ways) are eliminated. Comments often constitute the programmer's "notes to self" during the develop process, and, as a result, this valuable information is lost.

When a program is assembled or compiled, the assembly or compilation program can introduce strings and signatures into the code. Obviously, these sections of code must be identified and eliminated from consideration when we are trying to determine authorship of the program. (Occasionally, in virus research, compiler-introduced strings were mistakenly taken as unique and, therefore, used as signature strings for scanning programs. The antivirus scanners that used such strings would generate large numbers of false positive alarms as the strings were found in any programs that had been compiled from those languages.)

An additional concern is that compilers frequently optimize the code in some way, and this process may eliminate or confuse some parts of the characteristic signature of a given author.

As previously noted, other utilities besides compilers may be part of the program generation process. These utilities may also introduce signatures into the code, and these signatures must be taken into account. In addition, computer aided software engineering (CASE) tools and even programming environments (such as specialized editors directly associated with compilers) can influence the design and structure of programs. On the other hand, these various characteristics and signatures, if properly distinguished, can help identify a programmer or group, given a record of the use of specific sets of tools.

The source code that we receive with interpreted language programs does, generally, contain the comments (if the author made any, and did not eliminate them before releasing the program). We usually are not faced with compiler-introduced signatures, although a number of programming environments for interpreted languages may introduce comments or bias the use of certain types of programming styles or structures. However, the major concern with interpreted source code is that, particularly in regard to viruses and other widely distributed programs, the availability of the source means that a number of people have the opportunity to make minor variations to the program. This is easy to do when you have the source code, and interpreted languages tend to be simple, and are therefore within the programming skill level of a much wider group.

Finally, Already, the Tools

The first tool used in forensic research is obvious enough that most ignore it: a computer. I am not just being sarcastic at this point. A great deal of information can be obtained by noting the behavior and operation of the program under study when it is running. First, we can merely observe what the program does, in gross terms, as it runs. Then we can perform more detailed or low-level studies: Are attempts made to access specific areas of memory? Are calls being made to specific resources? Are attempts being made to contact other computers via a network, particularly the Internet? Then again, we can attempt to treat the program like a black box, and see what happens when we prod at it in various ways. (Of course, when dealing with malware, it is important to take precautions. If the first thing the program tries to do is to overwrite the hard disk, the information obtained can be limited.)

The next tool is the good old-fashioned hex editor. Used for displaying the content of binary files (in hexadecimal format, and usually also with those bytes that could be displayed in ASCII running parallel down the side), hex editors can help us find a number of interesting items that might be in the code.

The first items to look for are any strings of actual text. There tends to be a lot of text in programs. Some strings may be text that might appear as messages on the screen. Obviously, any program that contains a string stating "ha ha luzer i just blue up yer d!sc" probably warrants further study. When dealing with malware, as strange as it may seem, the authors of malware are often very proud of their creations, and may also include copyright notices, instructions for use, and even personally identifiable information.

Another set of strings that may appear as text in programs are application programming interface (API) calls. Particularly in Windows-based software, API calls can be very common. Even if you are not familiar with the libraries being used, APIs generally have very explanatory names. If, for example, you view the code for something that is supposed to be a game, APIs that indicate calls to close, open, or monitor network ports would be somewhat suspicious. An additional class of identifiable information might be available here: If calls are made to contact entities on the Internet, we may find uniform resource locators (URLs) or even email addresses.

As well as API calls, we may be able to recognize some function calls, although this takes a bit more practice. Programs use some printable characters (in fact, for Intel CPUs, it is quite possible to write programs using only printable characters), and some functions can be recognized by a particular string of ASCII characters. For example, in the old days of MS-DOS viruses, the string "PSQR" was one to watch for. It was related to a call by the program to "terminate and stay resident." Since few programs, in those days, needed to "go resident," such a call was an indication to look deeper.

Text strings may not appear in the program. In some cases, there may be no need for any. In other situations, malware authors may use simple forms of encryption to obfuscate messages. Generally, the encryption takes the form of a simple byte-by-byte XOR with a given byte value: For some reason, 2Fh seems to be quite popular. Cryptanalysis appropriate for simple substitution ciphers should be able to recover these text passages.

As there are assemblers and compilers for turning assembly and high-level languages into object code, so there are disassemblers and decompilers that do the reverse. Disassembly is easier than decompilation. However, note that disassemblers do not deal well with sections of text or data: They try to interpret the material as program code, with rather random results. In addition, malware authors frequently also encrypt sections of the code, specifically in order to frustrate attempts at disassembly. In this case, one must find the decryption routine, which must come in linear programming, prior to the encrypted section, and then use the function to decrypt the material before disassembly takes place.

Decompilers fare rather worse, and are a less mature technology in any case. Decompilers generally require assembly rather than object code as input, and usually do better if the language, and even version of the original compiler, can be determined. Decompilation is seldom fully successful, and most likely will produce some source code interspersed with sections of assembly code.

Another tool to use is a debugger, although those used in forensic programming differ from those used in high-level programming. Debuggers used in software forensics need to be able to control the execution of another program. Therefore, they need to act as a kind of software in-circuit emulator, allowing one operation at a time to proceed. The debugger should also have the ability to determine and display changes in memory and the CPU registers. The venerable

DEBUG, from MS-DOS systems, is able to perform a number of these functions, albeit in a very limited way with large programs, as well as function as a hex and sector editor and a disassembler.

Software Forensic Technologies and Practices

There are a variety of ways to look at code to obtain information and evidence. The most obvious, of course, is to look for text, functions, and other items in the content of the software. There are ways to look for patterns, not initially obvious, which are independent of the content of the program.

Content Analysis

We can use analysis of text and code to find sequences of messages and trace influences. In material that is copied from an original, the overall structure and composition is usually unchanged, but imitators have a penchant for adding details and embellishments.

The syntax of text tends to be characteristic. Does the author always use simple sentences? Always use compound sentences? Have a specific preference when a mix of forms is used? Syntactical patterns have been used in programs that detect plagiarism in written papers. The same kind of analysis can be applied to source code for programs, finding identity between the overall structure of code even when functional units are not considered. A number of such plagiarism detection programs are available, and the methods that they use can assist with this type of forensic study.

Of course, when considering the content of the text, most people consider characteristic use of vocabulary and phrases. This is frequently effective, but it usually relies on having a large set of samples to analyze. We also generally have to ensure that the texts cover the same or similar subjects, to avoid problems with disparate vocabularies in differing fields. Similar analysis can be applied to programs, using functional structures that provide analogs of vocabulary and assessing modules in the same way we read paragraphs.

Error Analysis

Errors in the material can be extremely helpful in our analysis, and should be identified for further study. In some of my early work published on the history of computer viruses, I made a mistake in the spelling of the name of one person involved in the creation of a specific program. Shortly thereafter, another person also published such a history. The histories were very similar, but that could be expected if two people both had access to the same sources. However, the second history also contained the error that I had made. The author of the second history, had he followed original reference materials, would not have made that error, thus indicating that the later text was a copy of my original.

The existence of errors in program code is problematic because certain types of mistakes will ensure that the program either does not compile or does not run. However, we may find that certain types of nonfatal errors, such as a failure to optimize various types of operations, may be characteristic of an individual or group.

Noncontent Analysis

A number of identifying attributes are available in order to build an electronic fingerprint of text. The same is true of program code. A specific method of finding such characteristics in text is called "cusum." Literary critics are quite used to talking about sentence length and structure as a characteristic of authorship, but other factors can be used as well.

Instead of looking at meanings or characteristic turns of phrase, cusum looks at combinations of statistical patterns in writing, patterns that the writer is probably unaware of using.

It may seem strange to use meaningless features as evidence. However, Richard Forsyth reported on studies and experiments that found short substrings of letter sequences can be effective in identifying authors of textual material. Even a relative count of the use of single letters can be characteristic of authors. Similar measures can probably be applied to program code, both source and object.

Additional Noncontent Indicators

Certain message formats may provide us with supplementary information. A number of Microsoft email systems include a data block with every message that is sent. To most readers, this block contains meaningless garbage. However, it may include a variety of information, such as part of the structure of the file system on the sender's machine, the sender's registered identity, programs in use, and so forth. In the case of material distributed by email, this information may be available to the forensic examiner.

Other programs may add information that can be used in our analysis. Microsoft's word processing program, Word, for example, is frequently used to create documents sent by email. Word documents include information about file system structure, the author's name (and possibly company), and a "global user ID." This ID was analyzed as evidence in the case of the Melissa virus. Microsoft Word can provide us with even more data: comments and "deleted" sections of text may be retained in Word files, and simply be marked as hidden to prevent them from being displayed. Simple utility tools can recover this information from the file itself. Some compilers create similar tables of data within the executable body of a program.

Legal Considerations

There will be differences in the permissibility of software forensic evidence, depending on the legal system that has jurisdiction over the crime. Admissibility of computer records may vary from system to system: Some legal systems will consider it hearsay and require higher standards for acceptance of it as evidence. Jurisdiction, as with any situation that deals with possible network involvement, may be a problem as well.

With respect to jurisdiction, of course, what may be considered a crime in one location may not be in another. Canadian law, for example, notes that anyone who, without authorization, modifies data or "causes" it to be modified, is guilty of an offense. Therefore, if we can, through software forensics, demonstrate that the intent of the program was to create data modification (possibly among other things) and to gain access to systems without active user involvement, then we have a case with regard to computer viruses. If we can, in addition, reveal a link to a specific individual as the author of the program, we can make a case against that person. Other jurisdictions may not have the same wording in law, and so we may not be able to prosecute certain types of activity. When using software forensics with respect to intellectual property cases, note that a number of countries, such as Pakistan, do not have intellectual property laws.

As this book is being written (2003), the legal situation with regard to software forensics, particularly in the United States, is very confused. Certain laws intended for the protection of intellectual property may be used to prevent the examination of software. There is the case of a programmer from Russia who came to the United States to speak at a conference about a weakness in a security mechanism in a commercial software product. He was, in fact, arrested and held in custody under the provisions of a law intended to protect against theft of intellectual property or the publication of ways to circumvent protections. Eventually he was released and never did face the charges in court. It may be possible that authors of malicious software may challenge software forensic evidence on the basis that they hold copyright on their software and did not grant permission for the software to be examined.

Presentation in Court

Presentation of this kind of technical evidence in court can be problematic. Debates over DNA evidence as identification, and the acceptability of such evidence to nonspecialists, which describes most lawyers, judges, and juries, are directly relevant to this issue. The field of forensic linguistics is still developing, and experts may have to be judged individually. Findings and opinions may be dismissed by the court on the basis that the expert cannot prove sufficient knowledge, skill, experience, training, or education.

Content-based analysis may seem to be a more reasonable choice for presentation, but its use may backfire. While it may be "morally" convincing, content analysis may still lack specific proof and be dismissed as mere opinion. In the embroidery chart example we will see in Chapter 8, it is instantly apparent that the patterns are from the same source, but it takes time to determine specific features and reasons, and the sequence of the patterns.

With the use of the word forensics in the title, legal factors are bound to be a major consideration. We will examine more of them in Chapter 5.

Summary

The analysis of software for evidence is, in one sense, a very old field. Through the arena of stylistic analysis, it stretches back to Biblical criticism and the hundreds of years of research in that subject, to arguments made at some of the earliest church councils determining the Biblical canon. Plagiarism detection and intellectual property disputes are far from new phenomena.

At the same time, the tools involved in software forensics have seldom been tested in court. In this latter view, then, software forensics is a completely new field.

This chapter was intended to provide a basic outline of what software forensics is, and how it may be used and pursued. We will go on, then, to examine the various components of the process in more detail.

2

The Players—Hackers, Crackers, Phreaks, and Other Doodz

Because we may be using software forensics to attempt to identify authors of software, it may help to have a rough idea of the type of people we are looking for. Those who write malicious software, or attempt to distribute or resell commonly available commercial software, tend to belong to communities of like-minded individuals. Over the years, we have been able to glean ideas about the characteristics of this tribe. For this information, we are all indebted to researchers such as Sarah Gordon, Dorothy Denning, Ray Kaplan, and, more recently, the members of the Honeynet Project.

A couple of provisos: Whenever you deal with people, there will always be exceptions. There are those who seem to pursue security breaking from motives that are, if not exactly admirable, at least untainted by thoughts of commerce or attention. Indeed, we can't really say that all endeavors related to the creation of viral software or intrusion utilities are even illegal. While most of the activity involved in security breaking is highly repetitive, there are also those few who do come up with one or two original ideas, and experiment with them.

As another example of a deviation from a stereotype, most studies of those involved in security breaking activities have been done in western societies: Europe, North America, and Australia. Recently, groups have been quite visible in China. There are two major populations, the red guests, and the black, or terrible, guests. The black guests are apparently quite akin to Western groups, with a lack of cooperation, antiestablishment positions, and random activities. The red guests, on the other hand, seem to form very stable groups, are executives in technology companies, have links with the Chinese government, and run coordinated exercises. In this case, we have a very large group running completely contrary to the expected norms for the community, and this may be derived from the differing foundations of Eastern and Western social thought.

Therefore, we can't make blanket statements about all of those within such a community. However, as with almost any stereotypes, there are reasons for the characterizations presented here.

Particularly in doing forensic analysis, we need to beware of falling into mental traps occasioned by our own "profiles" of the adversary. If we get too caught up in any one idea, we are going to blind ourselves to important evidence, whether it be proof of innocence or guilt. While it is beneficial to have an idea of the attributes of the majority of the people we are studying, it is absolutely vital always to accept the possibility of exceptions.

Terminology

When dealing with the blackhat communities and products, and malicious software in particular, there is a good deal of specialized jargon that does have meaning, but tends to be thrown around rather carelessly. Please bear with me in this section, as I will be mentioning some of it before it gets rigorously defined. By the end of the chapter, all should be clearly revealed.

The perceptive may have noted that I have not, except in the title, used the term "hacker." This is because there is considerable controversy in regard to the use of the word. Originally, the term meant one who was skilled in the use of computer systems, particularly if that skill was acquired in an exploratory manner. The usage applied to all aspects of the technology, whether hardware or software. In fact, it came to be extended to all forms of expertise: A hacker was a master of his (or her) craft, and the term was roughly equivalent to wizard or guru. Those who pursued this level of proficiency were usually those who were more than a little obsessed with it and therefore considered what the rest of the world sees as social skills to be, at best, inconsequential. Therefore, they gained a reputation for being uncommunicative and disdainful of notions of property and propriety, despite the fact that various forms of "hacker ethic" generally promoted the education of others and injunctions not to damage systems or data.

Later, the term came to be applied, usually by the media, to skilled or unskilled who break security systems. Originally, there may have been some merit in this usage. When the ability to communicate with computers at all was an arcane art, proficiency in connecting to them without authorization and getting them to perform for you was only acquired with patient investigation. However, as modems became more common, and as tricks for getting around access controls were distributed through bulletin boards, the level of skill required dropped significantly. It is easy to see why those who were trying to break into computers encouraged people to call them hackers: They assumed a mantle of mastery and superiority by virtue of a limited, though not really special, knowledge. In fact, a number of the members of the community came to be known as "wannabes" in their attempt to "want-to-be" seen as possessing skills that they did not, in reality, have.

Actually, you can determine a person's level of technical expertise by how he uses the term. Someone who uses hacker as meaning an expert is someone who generally does advanced technical work. Someone who uses hacker as a "bad guy" may have a technical background of some type, or a technical job, but usually is nowhere near the cutting edge.

So what do we call people who are breaking into computers or writing malicious software? An attempt was made some years ago to rehabilitate the term hacker, and to call those who tried to break, or crack, security systems, "crackers." Unfortunately, this attempt never did succeed with the general public, and there is a problem of confusion with those who break anticopying technologies on commercial software, who are also known as crackers.

In an attempt to avoid debates about "good" hackers versus "bad" hackers versus "crackers" versus "phone phreaks" versus "virus writers" versus "vxers" (and we will discuss the segmentation of the dark side population shortly), the security community has taken to describing those who either attempt to break into computer systems without prior authorization, or who explore security primarily from an attack perspective, as "blackhats." The term originates from the genre of old American western movies where the "good guys" always wore white hats and the "bad guys" always wore black. By a fairly automatic extension, those who attempt to explore security solely from the perspective of defense are the whitehats. (And, of course, with the world of computer security being convoluted, anyone who seems to sail fairly close to the line is known as a grayhat.)

Once again, I need to repeat my earlier point in regard to assuming too much. The term "blackhat" is a label of convenience for describing a broad class of activities and individuals. Not all people involved in blackhat groups are performing illegal activities. A series of security seminars has taken to using the term "Blackhat Conference." In regard to hiring those who have done computer intrusions to perform penetration tests of security, there is now discussion of "ethical blackhats." Therefore, it is safest to bear in mind that the term is most frequently used in regard to a perspective on systems and security, and to avoid dealing with moral judgments at this point.

Types of Blackhats

The blackhat community is extremely fragmented. Not only are there different groups, often at odds with each other, but the types of activities also differ. Despite the omnicompetent evil genii portrayed in fiction about "hackers," there is a great deal of specialization in the real blackhat groups, and those from one clique seldom do much exploration in the other fields.

Some are trying to break into or intrude on computer systems or networks. These are the ones who most frequently are given the hacker sobriquet, and are usually referred to as "crackers" (or system crackers, to distinguish them from the software piracy-type crackers) by the security community. Despite the general public reputation, few of these people do any programming, or create any sort of software, malicious or otherwise. There are a limited number of system crackers who do analyze software, and particularly system software, for weaknesses, and who then write exploit tools to automate the process of breaking in. However, these tools are, generally speaking, not a major problem. They are specific to a given system and version, and, even if distributed and utilized, have a very limited lifespan. If a particular vulnerability is widely exploited, then it tends to become known and patched quickly.

Other blackhats specialize in gaining unauthorized use of telephone switches and systems, usually for their own aims and amusement, but possibly for the purpose of obtaining or even reselling phone service. Those interested in breaking into or otherwise manipulating the telephone system are referred to (and refer to themselves) as "phone phreaks," using the punning variant spelling. This is generally shortened to "phreaks" in common usage. (Variant spelling, and even the use of nonalphabetic characters, is a characteristic of most blackhat communities. The effect is to define the population of the group, separating those who know the jargon, and therefore belong, from those who do not. Thus, those within can see themselves as members of an elite club—but probably represent it as "leet" or "3!33t." Hence also the reference to "doodz" [dudes] in the title of this chapter.) The act of manipulating the phone system is known as "phreaking."

Some are primarily interested in damaging or corrupting files, particularly in public ways, such as defacing Web sites. This runs

HACKTIVISM

Hacktivism is a convenient label, but a poorly defined term. Hacktivism can be anything that the user, generally a journalist, defines. It can be writing a new utility and releasing the same with attached political or social advertising. It can be developing a new Web site to promote civil rights or social change. It can also be developing online direct actions against corporations or governments, through mechanisms using the Internet.

The Internet enables debate or action on many issues. When we understand how people are using this new medium, we see a variety of social activities beyond online shopping and swapping pictures of family pets. For the online activists, such as the "electrohippies," understanding how different groups perceive the Internet is the first step in comprehending that these groups feel they are developing, or influencing, a new online consciousness that can create a new environment for realizing societal change, potentially globally.

contrary to versions of the "hacker ethic," because most of the documents identified as such contain some kind of "do no harm" provisions. However, many system crackers operate primarily on an ego drive, and need to have some way to prove an intrusion and keep score. In addition, a more recent attempt to prove the value of system breaking as a means of social protest, known as "hacktivism," uses the defacement of targeted Web sites as a vehicle for publicity and activism.

At its root, hacktivism is seeking to use advanced knowledge of IT systems to change the way people use and relate to computer hardware or computer networks, as well as each other.

A great many of the blackhats in general, and probably the largest majority, really have very little idea of the technology that they are using, having obtained packaged programs or scripts, and they are operating them without really understanding the functions or situations appropriate for their use. The vast majority of intrusion activity on the Internet arises from these "script kiddies" who have obtained utilities and scripts and simply launch attacks against random addresses. It is difficult to say that such attacks are even malicious. They are certainly thoughtless, and the time and resources necessary to deal with them are a drain on the resources of both institutions and individuals.

Those who create programs of any type, whether utilities or malware, are actually relatively rare. A number do make slight modifications to the creations of others, usually functionally insignificant changes to viruses, which are widely available because of their reproductive function. Thus, there are the vast, and usually closely related virus "families," and the phenomenon that when a new type of exploit tool arrives on the scene, it is quickly followed by a half-dozen extremely similar programs. The inordinately repetitive and noncreative nature of most of this programmed material may explain the contempt in which virus writers are held by large numbers of other blackhats. The production of viruses is seen, correctly, as a rather trivial exercise, rather than proof of programming skill.

There are, of course, those who are preoccupied in making illegal copies of commercial software. The "warez doodz" are generally most interested in collecting, and sometimes redistributing, such packages. A few, though, specialize in the analysis necessary to break systems designed to prevent just such copying. In some cases, these crackers also produce software dedicated to automating the copying or registering of commercial software.

And, at every level, there are those who "wannabe" more respected in the blackhat community, but lack even those skills.

Motivations and Rationales

It may be important to examine the commonly presented justifications for blackhat activity. There are two reasons for this study. First, this examination demonstrates something of the mindset and philosophy of the members of the community, and such a philosophy can sometimes be evident in programming style. Second, some of these justifications may be presented, quite seriously, as arguments against the activity of software forensics in general.

One of the most frequently attempted justifications of blackhat activity of all kinds is that it is protected under the concept of freedom of speech. Leaving aside the issue of whether free speech is a universal right, we also have to ignore for the moment the fact that most blackhat activity does not involve programming. In addition, we still have to ask whether programming is, or is not, speech. Speech generally does not involve other people, and when it does, such as in the case of yelling "Fire!" in a crowded theater or producing hate propaganda, it often is not protected. Programs may be used to express a message or idea of some sort, even beyond the text that such a system may carry or present. However, the determination of whether code actually constitutes speech can be extremely difficult, and has been decided both ways when presented before the courts. The concept of "artistic merit," which is usually considered in such cases, is unlikely to support the blackhat argument in terms of its usual products. In the case where the blackhat individual is not the author of the software, such as where attack scripts are being utilized or pre-existing viruses are being released, the protection of free speech is even more tenuous.

The freedom of speech argument resonates strongly, not only with society at large, but particularly with the blackhat community. Noting the self-identification with the original hackers, we frequently see statements such as "information wants to be free." There is a large measure of curiosity in the blackhat community, as an important characteristic. Among those who have been charged with computer trespass, a frequent claim is that the intruder "only wanted to know." While far from any justification, this motivation is probably true in a great many cases. The search for additional purposes is likely a waste of effort and a distraction from analysis of the real situation.

A second bid at vindication of security breaking activities is simply "because we can." Although the shallowness of this argument tends to prompt a sarcastic response from security or law enforcement personnel, we should note that the prevalence of this reasoning does make a very strong point about the anarchic nature and mindset of the blackhat community.

The idea of freedom itself is an important one. Note the competitive nature and divisiveness of the blackhat population overall. Note the tendency of blackhats to be loners and undersocialized. Freedom, in its most anarchic form, is an attribute of blackhats. Cooperation between individuals is rare, and between groups, exceptional. Therefore, evidence of multiple authors is frequently also an indication that code has been copied from another source.

Many individuals who practice system violation activity explain themselves on the basis that they are following in the footsteps of the old-time hackers, who explored and discovered the capabilities of early computing devices. This flies in the face of the reality of the current level of blackhat endeavors: The few instances that are not absolutely repetitive are generally slavishly derivative. Even if we ignore the fact that most "cracking" exercises amount to no more than "knocking on doors," we still have to ask what the objective of these explorations is, which usually cannot be clearly articulated, and look at the eventual result, which has not, to date, been anything significant.

However, in addition to the curiosity factor noted earlier, this does point out an important trait in blackhats. Ego drive is an extremely strong motivation. Therefore, we do not have to look for additional reasons, such as a profit motive, to explain activities. "I can do it," is quite sufficient. Searching for an idea of who would profit from a given operation is probably fruitless.

Yet another justification for blackhat activities is stated to be educational. As one who has been involved in education and training, as well as reviewing, for a great many years, I would be very sympathetic to this argument—if it had any basis. Even considering *2600* magazine, which can most charitably be described as the best of a bad lot, one is hard pressed to say anything positive about the writing quality, research, originality, or even such basics as sticking to the topic. When one turns to *phrack*, *40Hex*, and the myriad others of the "zine" ilk, the caliber runs steadily downhill. Even articles addressing simple penetration testing generally state only that systems are weak (we already knew that, thanks), and say nothing about strengthening them.

The educational activities, therefore, tend to be rather thinly veiled boasting. This characteristic may render all study of software forensics moot: Most individuals convicted of malware or security penetration offenses have been caught because of their own statements.

Hacker Manifestos and Other Documents

Please note that the texts reprinted here have been formatted for line length, but are otherwise unedited, and contain the original errors in grammar and spelling.

Almost from the invention of the computer there have been documents describing the characteristics and behavior of the skilled operator. The gist of a number of these has been compiled in the collection of materials known as "The Jargon File." This is an excellent source for gauging the mindset of those truly skilled with computers, and the type of position to which most individuals in the blackhat communities aspire. "The Jargon File" can be found at any number of Web sites, including http://www.elsewhere.org/jargon/html/index.html or http://info.astrian.net/jargon/.

From "The Jargon File," the entry on hacker ethics reads:

> "hacker ethic *n*. 1. The belief that information-sharing is a powerful positive good, and that it is an ethical duty of hackers to share their expertise by writing open-source and facilitating access to information and to computing resources wherever possible. 2. The belief that system-cracking for fun and exploration is ethically OK as long as the cracker commits no theft, vandalism, or breach of confidentiality."

In writings such as "Old and New Hacker Ethics," at http://cgi.fiu.edu/~mizrachs/hackethic.html, this is expanded to include items such as:

- Access to computers and hardware should be complete and total. (This is usually known as the hands-on imperative.)
- Information wants to be free. (This assertion has become almost a mantra and item of faith among various segments of the information technology world, both within and outside the blackhat community.)
- Mistrust authority and promote decentralization.
- Hackers should be judged by their hacking, not by irrelevant criteria such as race, age, sex, or position.
- You can create truth and beauty on a computer.
- Computers can change your life for the better.

Another set, found at http://www.crackinguniversity2000.it/alt-hacking-FAQ/15.html, includes the following points:

"Do not destroy/damage files unless it is absolutely necessary to cover your traces

"Notify the system administrator about any holes in security that you have found/exploited

"Patch any holes in security that you have found/exploited

"Do not break into a system for money. Do not steal money. Don't distribute information/software for money.

"Treat a machine that you break into as you would treat your own.

"Document/spread what you have learned"

Other prescriptive documents, outlining a level of moral activity, are similar to the "hacker ethic" found at http://www.datasync.com/~sotmesc/etc/ethics.html that reads as follows:

> "Hacking has Ethics you ask? Of course, though the media makes it seem like we are criminals, only a few of us are. I true hacker lives to know. A true hacker does not break into a system and delete its file system, plant and run viruses or try to destroy the data within. These however, seem to be the most known characterics of the "New Generation." These newbies, who most likely got a computer and Internet as a present, and only know the basic of whatever OS their system came with have a lot to learn.
>
> "They see a program, and they weigh the trouble of learning to use it well over how destructive it is. True

hackers use anonymous mail to cloak themselves, not send mail bombs. True hackers do not use Winnuke, or anyother DoS attack, unless it is to gain access to a server/network or in a act of self defense. Some of the newbies out there today, get a program like Winnuke simply to see how many people they can take out, if they'd research it a little more, they'd find out anyone with a patch is more then likely immune to Winnuke.

"True hackers, after the first week or so, release that if they keep asking for help, without even trying to find the answer will get flamed. These newbies, when finally gaining access to a system try to take it out. They try to employ programs like BO or Netbus, simply so they can use the term "Hacker" on there name. They are Wrong! More then likely they don't understand how the program even works, evidence alone on any hacking msg board, newbies asking questions without reading.

"Granted, I do not believe I should be called a true hacker yet. yet I know the difference between hacking and crashing, something more then likely the newbies will do. That or become a Warez Pirate, so they can be "k00l" or "313373" they have yet to understand what a hacker truly is, and most likely never will."

Currently, the most famous of the hacker credos is the "Hacker's Manifesto," written in 1986 by Loyd Blankenship under the pseudonym of The Mentor. It is reproduced in various forms around the World Wide Web, at sites such as http://manifestopost.com/famous/mentor.html or http://www.humanunderground.com/archive/hackermanifesto.html.

The text generally reads similarly to the following:

"Another one got caught today, it's all over the papers. 'Teenager Arrested in Computer Crime Scandal', 'Hacker Arrested after Bank Tampering.' 'Damn kids. They're all alike.' But did you, in your three-piece psychology and 1950s technobrain, ever take a look behind the eyes of the hacker? Did you ever wonder what made him tick, what forces shaped him, what may have molded him? I am a hacker, enter my world. Mine is a world that begins with school. I'm smarter than most of the other kids, this crap they teach us bores me. 'Damn underachiever. They're all alike.' I'm in junior high or high school. I've listened to teachers explain for the fifteenth time how to reduce a fraction. I understand it. 'No, Ms. Smith, I didn't show my work. I did it in my head.' 'Damn kid. Probably copied it. They're all alike.' I made a discovery today. I found a computer. Wait a second, this is cool. It does what I want it to. If it makes a mistake, it's because I screwed it up. Not because it doesn't like me, or feels threatened by me, or thinks I'm a smart ass, or doesn't like teaching and shouldn't be here. Damn kid. All he does is play games. They're all alike. And then it happened... a door opened to a world... rushing through the phone line like heroin through an addict's veins, an electronic pulse is sent out, a refuge from the day-to-day incompetencies is sought... a board is found. 'This is it... this is where I belong...' I know everyone here... even if I've never met them, never talked to them, may never hear from them again... I know you all... Damn kid. Tying up the phone line again. They're all alike... You bet your ass we're all alike... we've been spoon-fed baby food at school when we hungered for steak... the bits of meat that you did let slip through were prechewed and tasteless. We've been dominated by sadists, or ignored by the apathetic. The few that had something to teach found us willing pupils, but those few are like drops of water in the desert. This is our world now... the world of the electron and the switch, the beauty of the baud. We make use of a service already existing without paying for what could be dirt-cheap if it wasn't run by profiteering gluttons, and you call us criminals. We explore... and you call us criminals. We seek after knowledge... and you call us criminals. We exist without skin color, without nationality, without religious bias... and you call us criminals. You build atomic bombs, you wage wars, you murder, cheat, and lie to us and try to make us believe it's for our own good, yet we're the criminals. Yes, I am a criminal. My crime is that of curiosity. My crime is that of judging people by what they say and think, not what they look like. My crime is that of outsmarting you, something that you will never forgive me for. I am a hacker, and this is my manifesto. You may stop this individual, but you can't stop us all... after all, we're all alike."

General Characteristics

Blackhats, and particularly writers of malware and viruses, tend to be young, and almost invariably male. Despite occasional speculations on the addictive nature of "hacking," they usually "grow out of" the virus writing game after a few years. In terms of forensics, then, you are unlikely to find that the same author has continuously modified the same piece of malware over many years.

Why is the typical malware author young? Writers of fiction, and those in the media, tend to infer from this fact that technology is a young person's game, and that the elderly—those of the venerable age of say, 25—have lost the necessary mental acuity to produce viral code. When looking at the reality, however, this is far from the truth. With age comes experience, skill, and a library of past work. Therefore, programmers tend to become more productive in terms of the time taken to produce a given piece of code. Time is, though, an important factor. With age also comes a greater number of responsibilities and demands on time. Technical managers know that young programmers can be counted on to "pull all-nighters": older coders develop a slightly differing view of the importance of the work they are being asked to do.

In fact, malware authors are definitely those of whom it can be said that they should "get a life." Writing malicious code is not particularly difficult or sophisticated, but it does take a lot of time. Therefore, those involved in the practice are people who have nothing better to do. (In his thought-provoking essay, "Losing Your Voice on the Internet," James DiGiovanna points out that while virus writing is an attention-seeking behavior, it is inherently futile because the author's identity is seldom known to the general public, and the only importance of the virus itself is that it be eliminated.) In virus research, we have noted that virus authors almost universally mature beyond the malware game within a few years. The fact that most malware authors are also male has unfortunate implications for the importance of male, as opposed to female, roles in most societies.

Antivirus researchers tend to be dismissive of the technical abilities of virus writers. There are virus writers who write competent code; there are many more who do not. As noted earlier, the vast majority of malicious code is copied from earlier examples, with only very minor modifications. What variation is made tends to be cosmetic rather than functional: Often malware authors understand so little of the software they are working with that they dare not make changes to operational sections. In addition, malware is riddled with bugs of all types, indicating both carelessness and a general lack of skill.

The industry's lack of respect for the abilities of virus writers is well counterbalanced by the media, who continue to be fascinated by the mythical boy genius running rings round the incompetent antivirus workers. Malware authors like this cliché, too, and go to some lengths to encourage the stereotype.

Most of today's malware programmers gain access to a victim system by tricking the victim into executing malicious code. It is much easier to fool people than to identify possible exploits and find ways to effectively use them. Therefore, a preference for the "easiest" option is quite characteristic of malware programming.

Malware writers don't understand or prefer not to think about the consequences for other people, or they simply don't care. Recently, one researcher has speculated on the characteristics of the blackhat community in comparison to those of people who fall somewhere in the range between an admittedly ill-defined "normal" and those suffering from full-blown autism. Autistic individuals tend to perceive and interpret the world in an idiosyncratic manner.

Malware authors draw a false distinction between creating malicious software and distributing it. They eschew any responsibility for the damage caused by their creations. In particular, they believe it is the responsibility of the victim to defend him or herself from encroaching malware, not the responsibility of the creators to keep their handiwork away from systems other than their own. Targets and victims of attacks are typically dehumanized in blackhat writings, described as losers who do not deserve to own a computer. There is also projection and displacement of guilt, frequently expressed in terms justifying security breaking activities because a certain vendor makes poor quality software or because large corporations are doing bad things.

In self-reports from blackhats, a number of aspects are reported to be part of the thrill, including the act of vandalism itself, fighting authority, "matching wits" with the security or law enforcement communities, aggression (often arising out of resentment, and reinforced by the feeling of safety and power that is engendered by apparent anonymity), the ability to induce fear and panic in the media and the general public, and the "15 minutes of fame" as well as the recognition of peers. Malware writers tend to feel marginalized and unrecognized in normal society, so they feel a very strong sense of identity with the blackhat tribe, even while denigrating other members of that same community.

Blackhat Products

Most of the end result of blackhat activity consists of compromised systems, defaced Web pages, and pointlessly consumed bandwidth. Overall, this might be of interest to those investigating network forensics, but isn't of much use for us in software forensics. However, attack tools, distributed denial of service (DDoS) kits, trojans, viruses, worms, remote access trojans (RATs), and other forms of malware are.

We will, of course, want to find out as much as possible about what the specific piece of malware does. We also want to know about the author, if we can. Becoming familiar with the broad classes of malicious software can help point out, in general outline, the functions to look for. Knowing the class of malware may also help us to identify the author, because blackhats tend to be just as specialized as any other type of programmer.

It is sometimes difficult to make a hard and fast distinction between malware and bugs. For example, if a programmer left a buffer overflow in a system and it creates a loophole that can be used as a backdoor or a maintenance hook, did he or she do it deliberately? It may not be possible to answer this question with technical means, although we might be able to guess at it, given the relative ease of use of a given vulnerability.

It should be noted that malware is not just a collection of utilities for the attacker, although attack tools may still be legitimate items of study for software forensics. Once launched, malware can continue an attack without reference to the author or user, and in some cases, will expand the attack to other systems. There is a qualitative difference between malware and the attack tools, kits, or scripts that have to be under an attacker's control, and which are not considered to fall within the definition of malware.

There are a variety of types of malware. Even though functions can be combined, these types do have specific characteristics, and it can be important to keep the distinctions in mind. However, it should be noted that we are increasingly seeing convergence in malware. Viruses and trojans are being used to spread and plant RATs, and RATs are being used to install zombies. In some cases, hoax virus warnings are being used to spread viruses. Virus and trojan payloads may contain logic bombs and data diddlers.

Trojans, or trojan horse programs, are the largest class of malware. However, the term is subject to much confusion, particularly in relation to computer viruses. A trojan is a program that pretends to do one thing while performing another unwanted action. The extent of the "pretense" may vary greatly. Many of the early PC trojans relied merely on the filename and a description on a bulletin board. "Login" trojans, popular among university student mainframe users, mimicked the screen display and the prompts of the normal login program and could, in fact, pass the username and password along to the valid login program at the same time that they stole the user data. Some trojans may contain actual code that does what it is supposed to be doing while performing additional nasty acts that it does not tell you about.

A major component of trojan design is social engineering. Trojan programs are advertised (in some sense) as having some positive utility. The term *positive* can be in some dispute because a great many trojans promise pornography or access to pornography, and this still seems to be depressingly effective. However, other promises can be made as well. A recent email virus, in generating its messages, carried a list of a huge variety of subject lines, promising pornography, humor, virus information, an antivirus program, and information about abuse of the recipient's account. Sometimes the message is simply vague and relies on curiosity.

An additional confusion with viruses involves trojan horse programs that may be spread by email. In years past, a trojan program had to be posted on an electronic bulletin board system or a file archive site. Because of the static posting, a malicious program would soon be identified and eliminated. More recently, trojan programs have been distributed by mass email campaigns, by posting on Usenet newsgroup discussion groups, or through automated distribution agents (bots) on Internet relay chat (IRC) channels. Because source identification in these communications channels can be easily hidden, trojan programs can be redistributed in a number of disguises, and specific identification of a malicious program has become much more difficult.

Some data security writers consider that a virus is simply a specific example of the class of trojan horse programs. There is some validity to this usage because a virus is an unknown quantity that is hidden and transmitted along with a legitimate disk or program, and any program can be turned into a trojan by infecting it with a virus. However, the term "virus" more properly refers to the added, infectious code rather than the virus/target combination. Therefore, the term "trojan" refers to a deliberately misleading or modified program that does not reproduce itself.

In terms of programming, a trojan probably represents the simplest form of malware. It is, after all, trivial to write a program that will delete a file or format a disk. The only creativity involved relates to finding a cover story that hasn't been used so often that people will get suspicious.

A logic bomb is generally implanted in or coded as part of an application under development or maintenance. Unlike a RAT or trojan, it is difficult to implant a logic bomb after the fact, unless it is during program maintenance.

A trojan or a virus may contain a logic bomb as part of the payload. A logic bomb involves no reproduction and no particular social engineering.

A persistent legend in regard to logic bombs involves what is known as the salami scam. According to the story, this involves siphoning off small amounts of money (in some versions, fractions of a cent) credited to the account of the programmer over a very large number of transactions. Despite the fact that these stories appear in a number of computer security texts, the author has a standing challenge to anyone to come up with a documented case of such a scam. Over a period of eight years, the closest anyone has come is a story about a fast food clerk who diddled the display on a drive-through window, and collected an extra dime or quarter from most customers.

Also appearing more as a payload is a data diddler. This software deliberately corrupts data, generally by small increments over time. The slow and cumulative damage to the information may not be noticed for some time, and by the time it is remarked, previous backups will probably contain partially contaminated documents.

A hidden software or hardware mechanism that can be triggered to permit system protection mechanisms to be circumvented is known as a backdoor or trap door. The function will generally provide unusually high, or even full, access to the system either without an account or from a normally restricted account. It is activated in some innocent-appearing manner; for example, a key sequence at a terminal. Software developers often introduce backdoors in their code to enable them to re-enter the system and perform certain functions; this is known as a maintenance hook. The backdoor is sometimes left in a fully developed system either by design or accident. Backdoors can also be introduced into software by poor programming practices, such as the infamous buffer overflow error.

DDoS is a modified denial of service (DoS) attack. DoS attacks do not attempt to destroy or corrupt data, but attempt to use up a computing resource to the point where normal work cannot proceed. The structure of a DDoS attack requires a master computer to control the attack, a target of the attack, and a number of computers in the middle that the master computer uses to generate the attack. These computers in between the master and the target are variously called agents or clients, but are usually referred to as running "zombie" programs.

Again, note that DDoS programs are not viral, but checking for zombie software protects not only you and your system, but prevents attacks on others as well. It is, however, still in your best interest to ensure that no zombie programs are active on any of your machines. If your computers are used to launch an assault on some other system, you could be liable for damages.

Most people who actually launch DDoS attacks do not write their own software. Programs to control DDoS networks, slave or zombie programs, and packages to install the zombies are all available on the nets.

The authors of RATs would generally like to refer to these packages as remote administration tools to convey a sense of legitimacy.

All networking software can, in a sense, be considered remote access tools: We have file transfer sites and clients, World Wide Web servers and browsers, and terminal emulation software that allows a microcomputer user to logon to a distant computer and use it as if he or she were on site. The RATs considered to be in the malware camp tend to fall somewhere in the middle of the spectrum. Once a client such as Back Orifice, Netbus, Bionet, or SubSeven is installed on the target computer, the controlling computer is able to obtain information about the target computer. The master computer will be able to download files from, and upload files to, the target. The control computer will also be able to submit commands to the victim, which basically allows the distant operator to do pretty much anything to the prey. One other function is quite important: All of this activity goes on without the owner or operator of the targeted computer getting any alert.

When a RAT program has been run on a computer, it will install itself in such a way as to be active every time the computer is subsequently turned on. Information is sent back to the controlling computer (sometimes via an anonymous channel such as IRC) noting that the system is active. The user of the command computer is now able to explore the target, escalate access to other resources, and install other software, such as DDoS zombies, if so desired.

Once more, it should be noted that remote access tools are not viral. When the software is active, though, the master computer can submit commands to have the installation program sent on, via network transfer or email, to other machines.

Rootkits, containing software that can subvert or replace normal operating system software, have been around for some time. RATs differ from rootkits in that a working account must be either subverted or created on the target computer in order to use a rootkit. RATs, once installed by a virus or trojan, do not require access to an account. Once again, rootkits themselves may not be considered malware, although they can certainly be used for malicious purposes.

Other programs in this gray area between utilities and malware are sniffers. Sniffer packages essentially provide for eavesdropping on computer networks. Although they do not allow an avenue to information on machines, they do provide access to any data that flows between devices. This information can, of course, involve information about the structure and protections of the systems, including passwords and similar entry codes.

Pranks are very much a part of the computer culture. So much so that you can now buy commercially produced joke packages that allow you to perform "Stupid Mac (or PC, or Windows) Tricks." There are numberless pranks available as shareware. Some make the computer appear to insult the user; some use sound effects or voices; some use special visual effects. A fairly common thread running through most pranks is that the computer is, in some way, nonfunctional. Many pretend to have detected some kind of fault in the computer (and some pretend to rectify such faults, possibly making things worse). One entry in the virus field is PARASCAN, the paranoid scanner. It pretends to find large numbers of infected files, although it doesn't actually check for any infections.

Generally speaking, pranks that create some kind of announcement are not viral, and viruses that generate a screen or audio display are rare. The distinction between jokes and trojans is harder to make, but pranks are intended for amusement. Joke programs may, of course, result in a denial of service if people find the prank message frightening.

One specific type of joke is the "Easter egg," a function hidden in a program, and generally accessible only by some arcane sequence of commands. These may be seen as harmless, but note that they do consume resources, even if only disk space, and also make the task of ensuring program integrity very much more difficult. The presence of an Easter egg will definitely have an impact on software forensics, on the one hand increasing the volume of material that must be assessed, and on the other hand providing a potentially larger sample for comparison or statistical purposes.

Other Products

As noted, the activities of the blackhat community are primarily of interest when we are considering malicious software of various types. Software forensics may be used in a number of other cases, particularly in regard to intellectual property. In the case of stolen software, financial reward may be the sole consideration, and a random opportunity the only means. The perpetrator may not have any characteristics in common with the blackhat community.

In some situations, the theft of software may have a relation to the "warez" group noted earlier. However, it is unlikely that these anarchic individuals could put together a company selling commercial software, particularly the mass market variety. The likelihood is somewhat greater in relation to specialized niche software because corporate structure, distribution systems, and mass marketing has less of a role to play.

There is, of course, always the possibility of the theft of a certain piece of commercial code for inclusion in, say, an open source software project. This type of activity is more likely to be motivated by the same kind of ego drive we have noted is very important in blackhat circles. However, open source devotees are likely to spot this type of theft themselves, and will almost certainly reject any such donations in favor of home-grown versions.

Summary

A significant fraction of the work that software forensics may be called on to examine will have to do with malicious software, and other products of the blackhat communities. Therefore, without blinding ourselves to other possibilities, it is good to have a rough idea of the stereotypical characteristics we might expect to find in the authors of these programs. The attributes may be apparent in various aspects of the programming we are called on to examine.

The blackhat communities may be united in aspect, but they are very much divided in aims, activities, and nominal skills. Having an understanding of the various types of groups can be useful in regard to identifying the cultural influences that we observe when dealing with a specific piece of software.

It is important to understand that blackhat motivations may be substantially different from those found in other types of illicit activity. For those who are used to dealing with primarily profit-driven criminal behavior, the predominately ego-driven blackhat performance may be difficult to comprehend.

The various types of blackhat-produced software will come from different groups, with differing aims and skill sets. Therefore, it will be useful to have a realistic grasp of the multiplicity of types of software and the objectives behind them.

Having outlined this one broad class of objects of study, and the people behind it, we now turn to the more general description of the production of software code and the tools we will be using to examine it.

3

Software Code and Analysis Tools

I need to start by making a point that will be completely and simplistically obvious to those familiar with machine language programming—and totally bizarre to those who are not.

Machine language does not consist of the type of commands we see in higher-level languages. Higher-level languages use words from normal human languages, and so, while a given program probably looks odd to the nonprogrammer, nevertheless we see recognizable words such as "print," "if," "load," "case," and so forth, which give us some indication of what might be going on in the program. This is not true of machine language.

Machine language is, as we frequently say of other aspects of computing and data communications, all just ones and zeroes. The patterns of ones and zeroes are directions to the computer. The directive patterns, called opcodes, are the actual commands that the computer uses. Opcodes are very short, in most desktop microcomputers generally only a single byte (eight bits) in length, or possibly two. Opcodes may also have a byte or two of data associated with them, but the entire string of command and argument is usually no more than four bytes, or 32 bits, altogether.

Almost all computers in use today are based on what is termed von Neumann architecture (named after John von Neumann). One of the fundamental aspects of von Neumann architecture is that there is no inherent difference between data and programming in the memory of the computer. Therefore, in isolation, we cannot tell whether the pattern 4Eh (00101110) is the letter "N" or a decrement opcode. Similarly, the pattern 72h (01110010) may be the letter "r" or the first byte of the "jump if below" opcode. Therefore, when we look at the contents of a program file, as we do in Figure 3.1, we will be faced with an initially confusing agglomeration of random letters and symbols, and incomprehensible garbage.

```
-d ds:100 11f
B8 19 06 BA CF 03 05 FA-0A 3B 06 02 00 72 1B B4
  .........;...r..
09 BA 18 01 CD 21 CD 20-4E 6F 74 20 65 6E 6F 75
  .....!. Not enou

-u ds:100 11f
0AEA:0100 B81906        MOV     AX,0619
0AEA:0103 BACF03        MOV     DX,03CF
0AEA:0106 05FA0A        ADD     AX,0AFA
0AEA:0109 3B060200      CMP     AX,[0002]
0AEA:010D 721B          JB      012A
0AEA:010F B409          MOV     AH,09
0AEA:0111 BA1801        MOV     DX,0118
0AEA:0114 CD21          INT     21
0AEA:0116 CD20          INT     20
0AEA:0118 4E            DEC     SI
0AEA:0119 6F            DB      6F
0AEA:011A 7420          JZ      013C
0AEA:011C 65            DB      65
0AEA:011D 6E            DB      6E
0AEA:011E 6F            DB      6F
0AEA:011F 7567          JNZ     0188
```

■ **Figure 3.1** *Display of the same section of a program file, first as data, and then as an assembly language listing.*

Ultimately, understanding this chaotic blizzard of symbols will be of the greatest use to software forensic specialists. However, there are other objects for forensic study. Source code may be available, particularly in cases where we are dealing with script, macro, or other interpreted programming. To explain some of those objects, we need to examine the process of programming itself.

The Programming Process

In the beginning, programmers created object (or machine, or binary) files directly. The operating instructions (opcodes) for the computer, and any necessary arguments or data, were presented to the machine in the form that was needed to get it to process properly. Assembly language was produced to help with this process. Although there is a fairly direct correspondence between the assembly mnemonics and specific opcodes, at least the assembly files are formatted in a way that is relatively easy for humans to read, rather than being strings of hexadecimal or binary numbers. Notice in the second part of Figure 3.1 a column of codes that might almost be words: MOV (move), CMP (compare), DEC (decrement), and ADD (I hope I don't have to expand on that one). Assembly language added these mnemonics because "MOV to register AX" makes more sense to a programmer than simply B8h or 10111000. An assembler program also takes care of details regarding addressing in memory so that every time a minor change is made to a program, all the memory references and locations do not have to be manually changed.

With the advent of high, or at least higher, level languages, programming language systems split into two types. High-level languages are those where the source code is somewhat more comprehensible to people. Those who work with C or APL may dispute this assertion, of course. The much maligned COBOL is possibly the best example: As you can see in Figure 3.2, the general structure of a COBOL program should be evident from the source code, even for those not trained in the language.

Compiled languages involve two separate processes before a program is ready for execution. The application must be programmed in the source (the text or human readable) code, and then the source must be compiled into object code that the computer can understand: the strings of opcodes. Those who actually do programming will know that I am radically simplifying a process that generally involves linkers and a number of other utilities, but the point is that the source code for languages like Fortran and Modula cannot be run directly: It must be compiled first. (It is, of course, dangerous to make such statements: Undoubtedly some completist computer language historian will be able to identify Fortran and Modula interpreters of which I am totally unaware.)

```
    OPEN INPUT RESPONSE-FILE
         OUTPUT REPORT-FILE
    INITIALIZE SURVEY-RESPONSES
    PERFORM UNTIL NO-MORE-RECORDS
        READ RESPONSE-FILE
            AT END
                SET NO-MORE-RECORDS TO TRUE
            NOT AT END
                PERFORM 100-PROCESS-SURVEY
        END-READ
    END-PERFORM
begin.

     display "My parents went to Cape Cod and all they
got"
     display "for me was this crummy COBOL program!".
```

■ **Figure 3.2** *Two sections of code from different COBOL programs. Note that the intention of the program is reasonably clear, as opposed to Figure 3.1.*

Interpreted languages shorten the process. Once the program has been written, it can be run with the help of the interpreter. The interpreter translates the source code into object code "on the fly," rendering it into a form that the computer can use. There is a cost in performance and speed for this convenience: Compiled programs are "native" or natural for the central processing unit (CPU) to use directly (with some mediation from the operating system), and so run considerably faster. In addition, compilers tend to perform some level of optimization on the programs, choosing the best set of functions for a given situation.

However, interpreted languages have an additional advantage: Because the language is translated on the machine where the program is being run, a given interpreted program can be run on a variety of different computers, as long as an interpreter for that language is available. Scripting languages, used on a variety of platforms, are of this type. JavaScript applets, like the example in Figure 3.3, may be embedded in Web pages and then run in browsers that support the language regardless of the underlying computer architecture or operating system. (JavaScript is probably a bad example to use when talking about cross-platform operation,

```
<html>
<head>
<title>
Adding input
</title>

<!-- This script writes three lines on the page -->
<script>
document.write ("Hello, ");
document.write ("class.<br>This line ");
document.write ("is written by the JavaScript in the
  header.");
document.write ("<br>but appears in the body of the
  page,");
</script>

<body>
<!-- The following line is HTML, giving a line break and
  text -->
<br>This line is the first line that is written by HTML
  itself.<p>
Notice that this is the last line that appears until
  after the new
input is obtained.<p>

<!-- This script asks for input in a new window -->
<!-- Note that window, like document, is an object with
  methods -->
<script>
// Note that within scripts we use C++ style comments
// Internet & W3 How to Program p. 213 ff

// We declare a variable, studentName
// Internet & W3 How to Program p. 211 ff
var studentName;
// Then we get some input
studentName = window.prompt ("What is your name?",
```

■ **Figure 3.3** *A JavaScript applet that will work in all browsers. Note that this script uses much more internal commenting than is usually the case.*

```
  "student name");

/* Although we can use C style
   multi-line comments */
//.Internet & W3 How to Program p. 213 ff
</script>

<!-- This script writes a single line of text -->
<script>
document.write ("Thank you for your input, " +
  studentName);
</script>

</body>
</html>
```

Figure 3.3 (continued) *A JavaScript applet that will work in all browsers. Note that this script uses much more internal commenting than is usually the case.*

because a given JavaScript program may not even run on a new version of the same software company's browser, let alone one from another vendor or for another platform. But it is supposed to work across platforms.)

As with most other technologies where two options are present, there are hybrid systems that attempt to provide the best of both worlds. Java, for example, "compiles" source code into a sort of pseudo-object code called byte-code. The byte-code is then processed by the interpreter (called the Java virtual machine [JVM]) for the CPU to run. Because the byte-code is already fairly close to object code, the interpretation process is much faster than for other interpreted languages. Because byte-code is still undergoing an interpretation, a given Java program will run on any machine that has a JVM.

(Java does have a provision for direct compilation into object code. So do a number of implementations for interpreted languages such as BASIC. I think the language designers do this purely in order that explanations such as this cannot make straightforward statements.)

The Products

What we get, for analysis, depends, of course, on how the program was developed. If it was machine language programming, assembler, or a compiled language, we get an object code file for analysis. In the case of assembler or compilation, we may also have a copy of the assembler or high-level language source code. If we have an interpreted language used for development, we have a copy of the source code of the program. (For the purposes of software forensic analysis, partially compiled objects such as Java byte-code can be subject to the same type of analysis as object code. Also, source code, where available, can be assessed in the same manner regardless of whether the language used was a compiler or an interpreter.)

However, the development system still has some ways to make our analytical task more difficult.

Complicating Factors

When a program is compiled or assembled, all comments (unless they are handled in special ways) are eliminated. Comments often constitute the programmer's "notes to self" during the development process, and therefore, this explanatory information is lost when the object code is produced.

When a program is assembled or compiled, the assembly or compilation program can, itself, introduce strings and signatures into the code. Obviously, these sections of code must be identified and eliminated from consideration when trying to determine authorship of the program. (Occasionally, in virus research, compiler introduced strings were mistakenly taken as unique, and therefore used as signature strings for scanning programs. The antivirus scanners that used such strings would generate large numbers of false positive alarms, as the strings were found in any programs that had been compiled from those languages.)

An additional concern is that compilers frequently optimize the code in some way, and this process may eliminate or confuse some parts of the characteristic signature of a given author. For example, a programmer might specify the value of a variable every time he or she uses it, for reasons of his or her own. If the value is always the same, an optimizing compiler will set the value only once, and eliminate the other references. Therefore, this characteristic signature may be lost. In fact, compilers sometimes actually change the structure of a program to optimize memory or processing time. Therefore, when analyzing the compiled object code, we may be misled into thinking that there is a characteristic style for a programmer that actually results from the use of a specific compiler, or we may be looking for a characteristic signature that may have been eliminated by the use of a compiler.

As previously noted, other utilities besides compilers may be part of the program generation process. These utilities may also introduce signatures into the code, and these signatures must be taken into account. In addition, computer aided software engineering (CASE) tools and even programming environments (such as specialized editors directly associated with compilers) can influence the design and structure of programs. On the other hand, these various characteristics and signatures, if properly classified, can help identify a programmer or group, given a record of the use of specific sets of tools.

The source code that we receive with interpreted language programs does, generally, contain the comments (if the author made any, and did not eliminate them before releasing the program). We usually are not faced with compiler-introduced signatures, although a number of programming environments for interpreted languages may introduce comments, or bias the use of certain types of programming styles or structures. However, the major concern with interpreted source code is that, particularly in regard to viruses and other widely distributed programs, the availability of the source means that a number of people have the opportunity to make minor variations to the program. This is easy to do when you have the source code. Interpreted languages tend to be simple, and are, therefore, within the programming skill level of a much wider group.

The Resulting Objects

As you can see from the development process, there are a variety of objects that we may be called on to analyze in software forensic work. We may be fortunate enough to have the full and original source code for a program, including the developer's own comments. Source code is very much like written text, and analysis may be very much simpler when we have it to hand. Examination of source code is generally referred to as code analysis.

There may be situations where we have the source code for scripted or interpreted programs, but in a reduced form. Typically, scripts will not have any type of comments embedded in the code, and may be fairly minimal in terms of programming. Scripts often deal with common functions or operations, and therefore, may be copied from standard sources and possibly adapted to local specifics. In such a situation, the source code nature of the scripting language makes direct analysis of the intention of the program potentially easier, but determination of authorship may be quite difficult.

It is likely that the largest class of objects to be analyzed will be object code files. We will likely have the compiled object code, which must be examined from the perspective of a skilled machine language programmer. We may or may not have additional materials, such as printed or online documentation, but if such records exist, we also have to determine whether we can trust them to accurately reflect the nature of the software under study.

There is one more twist that may need to be considered. In the case of object code, we may not always have a file of the software. Firmware, programming embedded in hardware devices, presents special problems for the analyst. There are also programs resident only in memory, such as network worms that make direct attacks on servers and never do write copies of themselves locally. The copying of both firmware and memory-resident programs must be done carefully, and the procedures need to be forensically friendly if the results are to be presented in court.

The Analytical Tools

The first tool used in forensic research is obvious enough that most ignore it: a computer. I am not simply being sarcastic at this point. A great deal of information can be obtained by noting the behavior and operation of the program under study when it is running. First, we can observe what the program does, in gross terms, as it runs. Then we can perform more detailed or low-level studies: Are attempts made to access specific areas of memory? Are calls being made to specific resources? Are attempts being made to contact other computers via a network, particularly the Internet? Then again, we can attempt to treat the program like a black box, and see what happens when we prod at it in various ways. (Of course, when dealing with malware, it is important to take precautions. If the first thing the program tries to do is to overwrite the hard disk, the information obtained can be limited.)

The next tool is the good old-fashioned hex editor. Used for displaying the content of binary files (in hexadecimal format, and usually also with those bytes that could be displayed in ASCII running parallel down the side), hex editors can help us find a number of interesting items that might be in the code.

There are a variety of hex editors, and everyone involved with low-level programming or computer forensics will have a favorite, such as the frhed program pictured in Figure 3.4. One that isn't often considered, but is almost universally available, is DEBUG, which is used to view the same "Bride" sample file in Figure 3.5. This program has a number of severe limitations, including a serious restriction on the size of files it can use, but I will use it as an example because it came packaged with the DOS operating system, and therefore is likely available in most established computer offices. In addition, DEBUG has the basic functions of the other tools that we will be discussing. The documentation for DEBUG stopped shipping with later versions of DOS, but there is a very complete tutorial for the program, prepared by Fran Golden, available at http://www.datainstitute.com/chap1-11.htm.

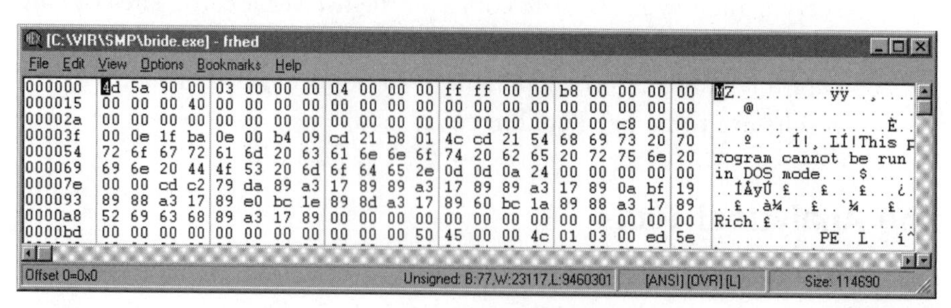

■ **Figure 3.4** *A view of the machine code of the file bride.exe using the frhed free hex editor for Windows.*

Generally, we will want to use a hex editor with a specific file. In the case of DEBUG, there are two ways to do this. The first is simply to start the DEBUG program with the name of the file as an argument, giving the command "debug filename.ext" to the system. However, you can establish the association once DEBUG has started by using the name command, which is the letter "n" (or "N": DEBUG will recognize either case). Therefore, within a DEBUG session, you can give the command:

```
n filename.ext
```

```
0E 1F BA 0E 00 B4 09 CD-21 B8 01 4C CD 21 54 68
  ........!..L.!Th
69 73 20 70 72 6F 67 72-61 6D 20 63 61 6E 6E 6F
  is program canno
74 20 62 65 20 72 75 6E-20 69 6E 20 44 4F 53 20
  t be run in DOS
6D 6F 64 65 2E 0D 0D 0A-24 00 00 00 00 00 00 00
  mode....$.......
CD C2 79 DA 89 A3 17 89-89 A3 17 89 89 A3 17 89
  ..y.............
0A BF 19 89 88 A3 17 89-E0 BC 1E 89 8D A3 17 89
  ................
60 BC 1A 89 88 A3 17 89-52 69 63 68 89 A3 17 89
  '.......Rich....
```

■ **Figure 3.5** *The same file displayed using the DEBUG utility. Note the same text strings are evident.*

to get DEBUG to work on a particular file.

DEBUG being a very simple utility, associating the filename is not enough. You must give a command to load the file, or a section of it, into memory. The load command is, appropriately, load or "l." Therefore, having associated with a filename, we then issue the l command in order to read the file into memory for further work and often assign it to a specific location in memory. However, one of the advantages of DEBUG as a hex editor is that it can, in fact, work in the absence of a filename. We can, if we know the structure of the disk, load sections of the disk that are not linked to any files. When we are not using filenames, the options for the load command are "L [memory address] [drive] [first sector to be loaded] [number of sectors to load]." For example:

```
l ds:100 0 0 1
```

translates as "load into memory location ds:100, drive A (0), sector zero, and load one sector." This is, in fact, the command to load the boot sector of a floppy disk in drive A. Using this form of the load command, we can access system and hidden sectors on the disk and parts of the disk that are used to store and manipulate the file system itself.

The last command needed for the hex editor function is to display or dump the information, the command "d." The command:

```
d ds:100 2ff
```

will dump the contents of the full boot sector that we just loaded (although the complete contents will also overfill the screen).

The first items that we may want to look for when examining files with a hex editor are any strings of actual text. There tends to be a lot of text in programs. Some strings may be text that might appear as messages on the screen. Obviously, any program that contains a string stating "ha ha luzer i just blue up yer d!sc" probably warrants further study. As strange as it may seem, the authors of malware are often very proud of their creations, and may also include copyright notices, instructions for use, and even personally identifiable information about themselves. If we are dealing with intellectual property cases, it is possible that the plagiarist may have left text identifying the code as the property of the original author unchanged in the body of the program.

There are specialized programs that will look for strings of text characters within object code, and these can save you time in finding such passages. An example is BinText, illustrated in Figure 3.6. Text strings may not be apparent in the program. In some cases, there may be no need for any. In other situations, malware authors may use forms of encryption to obfuscate messages. Generally, the encryption takes the form of a simple byte-by-byte XOR with a given byte value: For some reason, 2Fh seems to be quite popular. Cryptanalysis appropriate for rudimentary substitution ciphers should be able to recover these text passages. There are other utilities that automate the process. As you progress in the software forensic field, you will undoubtedly build a toolbox with a variety of favorite utilities along these lines.

A number of text strings are used as signatures in nontext files. For example, most DOS and Windows program files start with the characters "MZ" (or sometimes "ZM"). Graphic interchange format (GIF) files start with the letters "GIF" (usually "GIF89a"), bitmap graphics files begin with "BM," and Joint Photographic Experts Group (JIF) have "JFIF" six bytes from the origin of the file. A number of other standard signatures are not text, but can still be found with hex editors: Microsoft Office document and worksheet

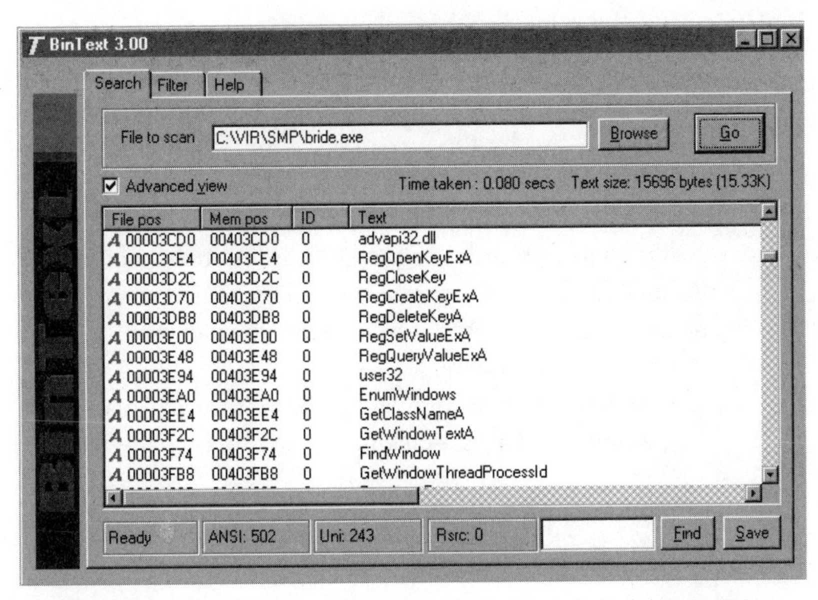

■ **Figure 3.6** *The BinText utility, finding text strings in the sample Bride program file.*

files start with the hexadecimal characters D0 CF 11 E0 A1 B1 E1. You can, in fact, use these signatures to find files inside files: If you look at a Microsoft PowerPoint show that contains JPEG images, you will see the JFIF signature at numerous places in the file, and you can therefore extract the images.

Another set of strings that may appear as text in programs are application programming interface (API) calls. Particularly in Windows-based software, API calls can be very common. Even if you are not familiar with the libraries being used, APIs generally have very explanatory names. If, for example, you view the code for something that is supposed to be a single-player game, APIs that indicate calls to close, open, or monitor network ports would be somewhat suspicious, as there would normally be no need for such a function in that kind of software. An additional class of identifiable information might be available here: If calls are made to contact entities on the Internet, we may find uniform resource locators (URLs) or even email addresses.

As well as API calls, we may be able to recognize some function calls, although this takes a bit more practice. Programs use some printable characters (in fact, for Intel CPUs, it is quite possible to write programs using only printable characters: see the "Printable

Programs" sidebar), and some functions can be recognized by a particular string of ASCII characters. For example, in the old days of MS-DOS viruses, the string "PSQR" was one to watch for. It was related to a call by the program to "terminate and stay resident." Since few programs, in those days, needed to "go resident," such a call was an indication to look deeper.

Of course, in order to pick out useful items, you need to know machine language. Once you start to recognize what functional elements look like in actual object code, you can employ this information in a variety of ways.

It is beyond the scope of this book to teach machine language programming. That is a task for a book all on its own. An excellent

PRINTABLE PROGRAMS

The characters you find in machine language (object code) programs can have any of the 256 possible values for a byte. However, not all of those values have a printable representation onscreen in a single character. (Indeed, some are specifically defined as nonprinting: Character 07h is the BEL, or bell character, and should sound a tone rather than generate anything on the screen. 00h is the NUL character, and should not display anything at all.) However, there is enough redundancy in the Intel opcode set that any type of program can be created using only those characters that do have a specific character representation. These printable programs formerly had uses in transmitting programs over email systems, before the technology of attachments was mastered.

One example, and a small program for disassembly practice, is the EICAR signature test file. This is a working program that uses only printable characters. It can be created by copying the following 68 characters:

```
X5O!P%@AP[4\PZX54(P^)7CC)7}$EICAR-STANDARD-ANTIVIRUS-TEST-FILE!$H+H*
```

exactly as shown, into a file, which can then be named with an executable extension, such as .COM or .EXE. More information on the file (and various downloadable versions) can be found at the European Institute for Computer Anti-Virus Research (EICAR) Web page at http://www.eicar.org/anti_virus_test_file.htm.

guide to start the process is *Assembly Language Step-by-Step*, by Jeff Duntemann. Duntemann provides everything necessary to get started, including an assembly language programming environment, or, for those not wanting to be restricted by yet another specific system, instructions on how to use DEBUG. (Yes, DEBUG again. It has an assemble command: By now you should be starting to realize that this would be "a".)

Returning to the initial tool—the computer—there are ways of getting more information from it. We can, of course, just run a program and see what happens on the screen or elsewhere. However, if we look deeper, we can obtain more data. We have noted that the dump command in DEBUG can be used to display the contents of a file that we have loaded. However, what we are really doing is displaying the contents of the computer's memory. Because the memory will be used by programs, examining the memory before and after, or even during the run of a program can give us a variety of insights. There are, of course, other programs besides DEBUG that

Learning Assembly Language with Emulators

Windows machines are not ideal platforms for those who are just starting to learn assembly programming. To fully understand how computers work at the most basic level, you need something simpler, where you can see every change that a command makes. Colleges and training institutions typically start machine language programmers off with simple devices consisting of a rudimentary CPU, a limited amount of memory (one kilobyte is considered ample), some input toggle switches, and an elementary display.

This kind of apparatus is very similar to the early hobbyist computers, such as the Altair and Imsai. With those ancient (in Internet years) machines, it was easy to see how the computer operated at a fundamental level. Modern computers have layers upon layers of systems, components, and utilities, all devoted to hiding the low-level computing processes from the user.

Fortunately, you can now get emulators of the Imsai (Figure 3.7) and the Altair (Figure 3.8). These are software versions of the machines, but allow you to enter and run programs and determine the state of the machine at every point.

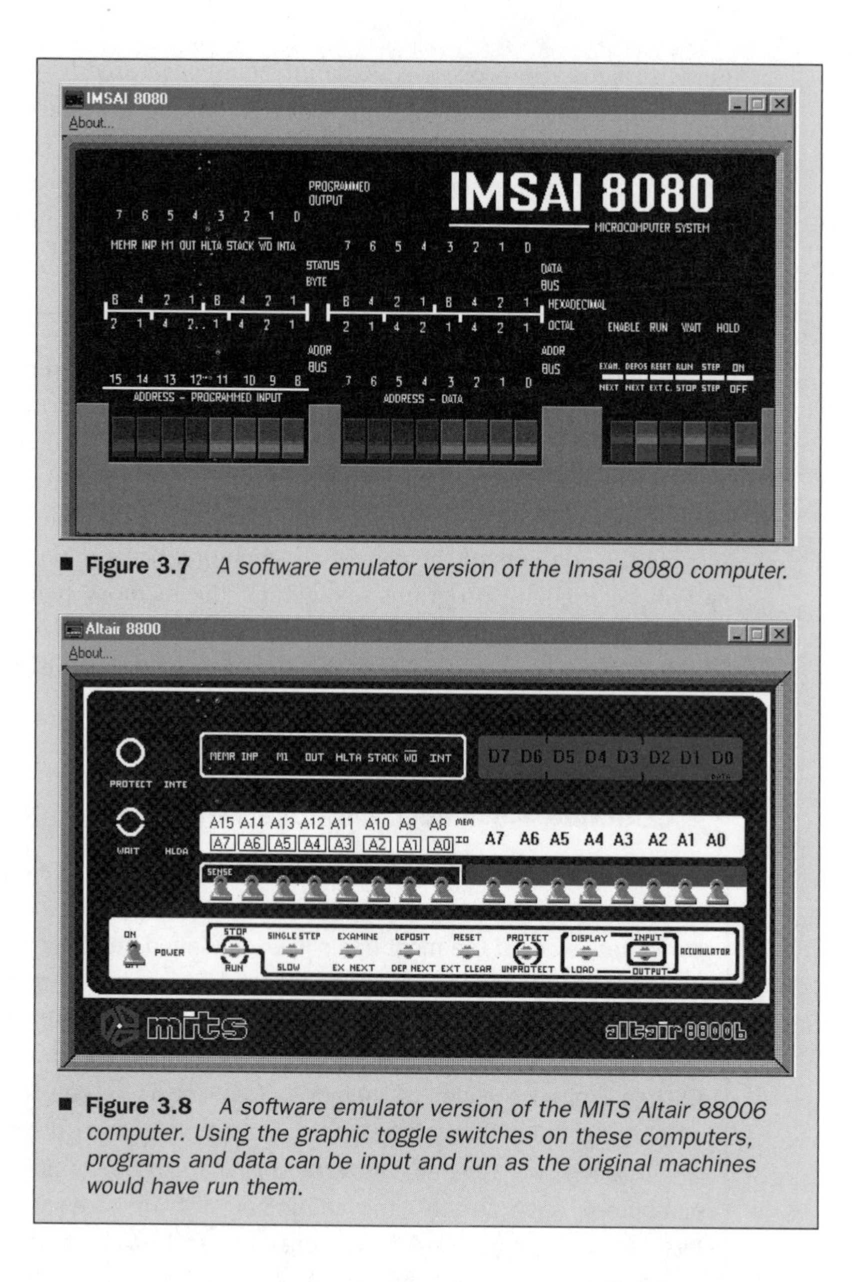

■ **Figure 3.7** *A software emulator version of the Imsai 8080 computer.*

■ **Figure 3.8** *A software emulator version of the MITS Altair 88006 computer. Using the graphic toggle switches on these computers, programs and data can be input and run as the original machines would have run them.*

can accomplish this type of task. Indeed, DEBUG's limitations would make it almost impossible to use as a memory examination tool in these days of massive memories, as compared with the 640K limit of DOS days. Other tools will be able to seek through memory and show us sections that have changed.

The random access memory, and even cache memory that we normally consider as being next to the heart of the computer, is really only convenient storage. Concerning the CPU, the only data that can actually be manipulated are the tiny amounts contained within the registers on the CPU itself. Once again, DEBUG (and other programs) comes to the rescue. The command "r" followed by suitable arguments will show us the contents of the registers themselves on the CPU. As with other forms of memory, it is important to watch the changes that occur. However, because the registers contain only one datum at a time, and effectively have no way to track historical values, it is only useful to view the registers if we can step through a program, one instruction at a time. And DEBUG has a trace command ("t") that does this.

Another tool to use is a debugger, although those used in forensic programming differ from those used in high-level programming. Debuggers used in software forensics must be able to control the execution of another program. Therefore, they need to act as a kind of software in-circuit emulator, allowing one operation at a time to proceed. A forensic level program should be intelligent and wide-ranging enough to be able to determine attempts to bypass its control of the computer. The debugger should have the ability to determine and display changes in memory and the CPU registers. The information returned by the debugger will be useless if malicious software is somehow able to modify it.

As there are assemblers and compilers for turning assembly and high-level languages into object code, so there are *disassemblers* and *decompilers* that do the reverse. Disassembly, or unassembly, as it is known to DEBUG ("u"), can be fairly straightforward. Because assembly is simply translating mnemonics into opcodes, opcodes can be reasonably accurately turned back into mnemonics. Of course, comments and other human-friendly but machine irrelevant pointers will be lost, but the program structure and function should be readily apparent. Using DEBUG, the command:

```
u ds:100 10f
```

given following our loading of the boot sector earlier, will disassemble the first f (in hexadecimal, 15 in decimal terms) bytes of the previously loaded boot sector.

With certain provisos, note that disassemblers do not deal well with sections of text or data. They try to interpret the material as program code, with rather random results. Remember from our earlier discussion that there is no inherent difference between data and code, and so other clues must be used to make a distinction. DEBUG is fairly basic in this department. It will simply try to convert everything from opcode to mnemonic. Other disassemblers are more intelligent, and try to determine what might be sections of data, best left alone. One highly regarded disassembly program is IDA Pro (Figures 3.9 and 3.10).

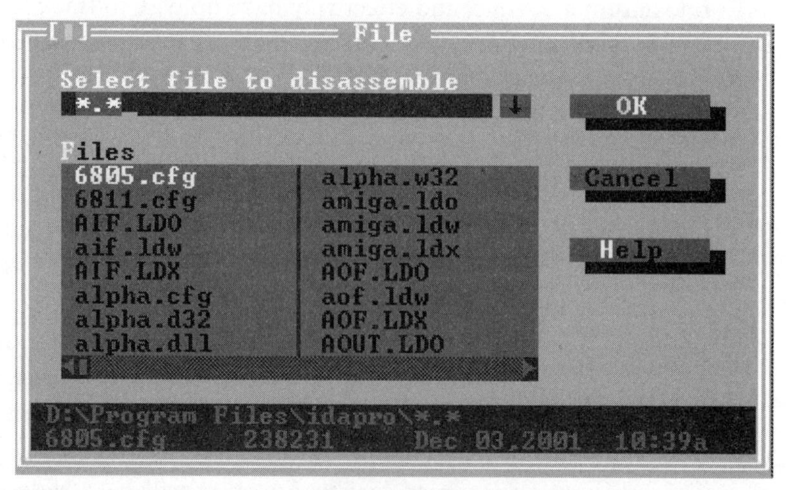

■ **Figure 3.9** *The IDA Pro for Windows disassembler in starting configuration.*

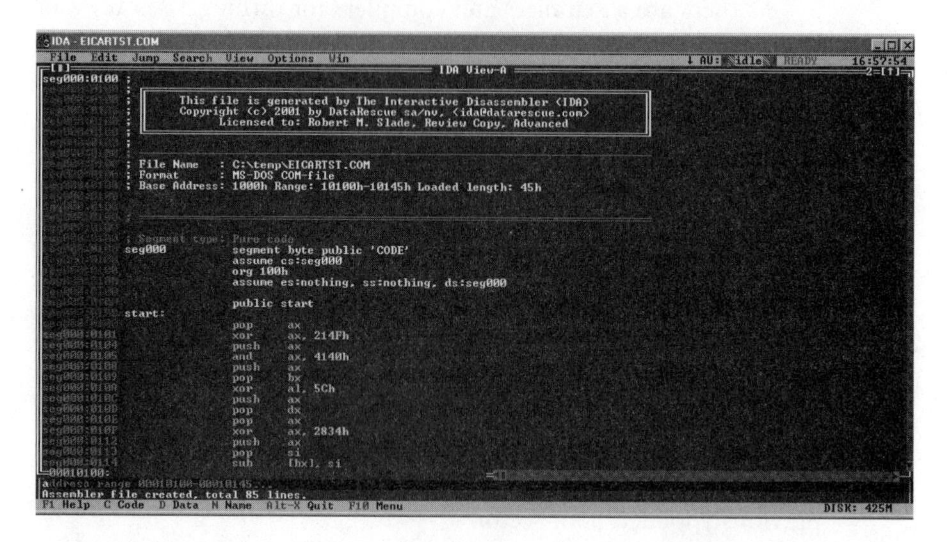

■ **Figure 3.10** *The IDA Pro for Windows disassembler performing a disassembly of the EICAR test file.*

This type of activity is hard enough, without people actively trying to make it worse. In addition to trying to hide text messages, malware authors frequently encrypt sections of the code, specifically to frustrate attempts at disassembly. In this case, one must find the decryption routine and then use that to decrypt the material before disassembly takes place. It is not necessarily difficult to find the decryption routine. Execution starts at the beginning of the program, and so stepping through the execution, including any jumps taken along the way, will inevitably start decryption before the main body of the program can be invoked. This is because the body must be decrypted before it can be run. Still, tracking down encryption is time-consuming and adds to the work.

Disassembly is easier than decompilation. Decompilers fare rather worse than disassemblers and are a less mature technology in any case. Decompilers generally require assembly, rather than object code as input, and usually do better if the language, and even version, of the original compiler can be determined in advance. Decompilation is seldom fully successful, and most likely will produce some source code interspersed with sections of assembly code. We will look at decompilers in more detail in the next chapter, dealing with advanced tools.

Forensic Tools

Although they are not specifically intended for software forensics, we should take note of the traditional computer forensics and data recovery tools. As we noted at the beginning of the chapter, there is no defining difference between data and code. With a few notable exceptions, software objects are going to be, or appear to be, standard data files. Therefore, any tools dealing with the recovery or preservation of data evidence can be applied to the recovery of software.

When forensically examining a computer, one of the first injunctions is to do a memory dump. This provides information about the current state of the machine and recent transactions. In terms of software forensic analysis of suspected malicious software, a memory dump may be the only means of obtaining copies of memory-resident malware such as network worms. There are both software and hardware tools for obtaining a memory dump. Hardware devices are probably superior in terms of obtaining the contents of memory without disturbing the memory structure itself, but may present compatibility problems with new forms of memory.

The second commandment for seizing a computer is to make multiple complete image copies of the hard drive, protect and preserve the original hard drive, and then perform any necessary analysis on one of the copies. In addition, cryptographically secure hashes should be taken of the full image and all relevant files. This is done to prevent any charges of tampering with the evidence during the course of the examination. In the case of software forensics, the preservation of the evidence, and proof that it has not been modified, is probably even more important. The disk image is also an important tool because it allows for exploration of slack space, "bad" sectors, normally unused locations, and other hiding places for malicious software.

Summary

This chapter has been a bare introduction to the process of programming that produces the items software forensics studies. We have also looked at the basic types of tools used in examination and analysis. The next chapter will describe more advanced and specialized tools for particular purposes.

4

Advanced Tools

The items described in this chapter are not commercial software packages: not yet, at any rate. The technologies examined here are primarily still in the research stage. In all likelihood, when this kind of software does become available, it will use a variety of the approaches that will be reviewed. This chapter is intended to present the concepts and note some of the work that is being done.

Decompilation

In chapter three, we described the process of compiling programs from source code to create object code files that the computer can run directly. Because object files are produced by compilation of the source code, it should be possible to decompile the object code and recover something very similar to the original source. This is analogous to the disassembly that we also noted in chapter three, where the object code is partially "explained" by recovering the mnemonics of the opcodes and recreating the more legible assembly language program.

Determination of the intent of a piece of unknown software, and particularly malware, will be much easier with source code than with the object code alone, but malware source code is not always available. Software forensic analysis is not the only time when such a program might be desired. Legacy code for older programs can sometimes run for years on systems, and when maintenance needs to be done, the original source code may not be available. Similarly, if the hardware is to be upgraded, porting of the software will need to be performed. Such porting is much easier if the source code can simply be recompiled for the new hardware. Decompilers could also be used to verify that the original compilation took place correctly. If the decompiled source is significantly different from the prototype, it could indicate a problem.

Ironically, decompilers may be of particular help in porting programs to new platforms requiring new high-level languages. Because object code will be very similar regardless of the programming language used to produce it, a given piece of object code may be decompiled into a number of higher level language "source" code programs. While this will be helpful for recovery of source code, and possibly analysis of program intent, in regard to plagiarism and authorship analysis, we must be careful that what we recover is close to the original. Therefore, decompilation programs may need to perform different types of analysis depending on whether they are simply recovering the structure and operations of the program, or attempting to determine the original language and possibly even version of the compiler that was used.

Machine code translation into assembly language with a disassembler is relatively easy. Because assembly language is a list of the most primitive instructions used by the computer, there is a correspondence between every opcode in the program and a given assembler mnemonic. However, there is not simply a direct correlation between every byte in the object code and an assembly command. Opcodes also require operand data (generally memory addresses), so the disassembler must know how many bytes of operand a given opcode needs to leave as data, rather than try to interpret them as opcodes. Then we need to identify any bytes or strings of actual data that are coded within the program. The ability to deal with these issues distinguishes basic disassemblers from the more professional models.

In a high-level language, many primitive instructions are joined together into a single command or statement. The language specification will indicate the pattern of opcodes that a given statement should produce. In the simplest of programming languages, this is accomplished with a type of library of preprogrammed modules. In more advanced development tools, particularly those using optimizing compilers, the patterns produced are more complex and depend on the structure of the rest of the program.

Decompilers have existed almost since the earliest high-level languages. Reverse engineering and compilation is still a relatively small area of research, possibly due to the complicated legal issues, because a reverse compilation tool may be abused in recovering code for proprietary programs and possibly plagiarizing material. Even before the recent and intricate UCITA and DCMA legislation in the United States (discussed further in Chapter 5), software vendors often made injunctions against reverse engineering as part of the user agreement.

Desquirr

David Eriksson has made a partial attempt at creating a decompiler program with a package called Desquirr, available at http://www.2good.com/software/desquirr/. His system is limited in a number of ways. For example, it uses the commercial tool Interactive Disassembler Pro (IDA Pro) to create assembly code from the object code, and then it performs the decompilation on the assembly source.

Reverse compilation can be accomplished in two different ways. Control flow analysis reviews the conditional and nonconditional jump instructions found in the object and assembly code and expresses them in terms of the high-level control constructs such as if/then/else, switch/case, for-loops, and while-loops. This allows you to obtain a view of the overall structure of the program. Data flow analysis, on the other hand, recovers expressions dealing with the actual program actions and functions. Desquirr concentrates on data flow analysis. Other programs (such as dcc, discussed in the next section of this chapter) have attempted to deal with control flow analysis. A commercial decompiler for forensic use would need to deal with both elements, as well as work directly from the object code without requiring outside disassembly.

Desquirr, and most other decompilers to date, have primarily dealt with the C programming language. C is actually a rather low-level language, as programming languages go. One of the characteristics of C is that most data structures must be defined within the program, which makes the decompilation of data structures quite problematic. Other languages, with more predefined data structures, may be easier to decompile in this regard.

An additional difficulty results from the use of optimizing compilers. In Chapter 7, we discuss the fact that there are always multiple ways to do things. Individual programmers find their own preferred methods. Optimizing compilers may look at a specific piece of programming and select another way to do the same thing, using fewer bytes of code, fewer machine cycles, or less space in memory. The original structure is therefore replaced with the optimized version. In a general analysis of program intent, this may not matter. However, it can have an impact on the analysis of authorship issues.

Dcc

Dcc is an older decompiler written by Cristina Cifuentes and available from http://www.it.uq.edu.au/groups/csm-old/dcc.html. As well as taking the structural control flow approach to decompilation, dcc is more of a stand-alone system. It deals with disassembly, analysis in an abstract format, and conversion to the high-level language (C, in the case of the existing version of dcc).

The results from dcc were used in work on the University of Queensland Binary Translator (UQBT, http://www.itee.uq.edu.au/~cristina/uqbt.html), which is pursuing the goal of automatic porting of object programs from one hardware platform to another. It also made a major contribution to the Boomerang open source project.

Boomerang

Boomerang, available from http://boomerang.sourceforge.net/, is the closest approach to a full decompiler. Unfortunately, it is still very much a work-in-progress. In fact, a recent addition to the package that may ultimately ensure more accurate and universal code analysis has actually degraded performance in the short term. Still, this project holds a lot of promise.

Plagiarism

Most of the work currently involved in the detection of plagiarism concentrates on identifying cheating on college assignments, particularly those involving programming. Given that most software forensic research is still in the academic phase, this emphasis is completely understandable. As a college and university instructor, I have always seen a few cases where one student will turn in a project that completely duplicates that of another. With the increasing availability of material on the Web, the standard programming assignments and programs that fulfill them are all available for downloading.

The use of software forensics to determine plagiarism in the legal arena will more likely revolve around intellectual property disputes. The programs that have been developed for the academic field will not be directly applicable, but the technologies, algorithms, and approaches will be.

The point should be made that plagiarism detection and authorship analysis are not always the same. When determining the author of a given piece, one must assume that we may only start with the piece itself and have to collect characteristics and other indicators that may be able to direct us to the author. In the case of plagiarism, we may be able to compare the piece in question with an existing body of work and test for matches between the suspect item and the original exemplars. The two approaches may use common technologies, and each can inform the other. The intents and purposes, though, are different, and should not be confused.

JPlag

JPlag, from the University of Karlsruhe and publicly available at http://www.jplag.de, is a program designed to detect plagiarized student Java programming assignments (Java PLAGiarism), although C, C++, and Scheme can also be considered. An interesting feature of the system is that it is available as a Web service: a fairly obvious design choice when you consider that the users may want to be able to compare submissions with material available on the Internet. The basic algorithm used, called Greedy String Tiling, is a favorite with researchers in plagiarism detection.

The main emphasis in JPlag appears to be detecting duplications in a set of similar programs: for example, a few hundred student assignments on the same problem. Since the problem is the same, and the teaching is the same, it is likely that the programs turned in by the students will be substantially similar. It will be that much harder to determine whether likenesses are due to the correspondence in the problem, or plagiarism.

JPlag takes a set of programs and compares pairs of them for a total similarity value and also regions of similarity. It generates a set of HTML pages that allow for exploring the similarities found.

JPlag converts each program into a set of abstract symbols, and then, where possible, covers (or "tiles") one such token string by substrings taken from the other (string tiling). The conversion of the source code into symbols is dependent on the language used in the source code, but the tiling is not. In future versions of the program, it may be possible to detect plagiarism even where the source code language has been changed.

The authors of JPlag note that earlier research into plagiarism detection often used the approach of comparing vectors or metrics. These technologies were of low sensitivity, and either missed too many duplicate entries or generated too many false alarms. This has an interesting relation to the debate (in forensic linguistics, which we discuss in Chapter 8) over the use of content or noncontent metrics. At first thought, it seems to be intuitively obvious that examining the content of a document allows us to more accurately determine whether copying has taken place, or who the author may be. Research increasingly indicates that this assumption is false, and that noncontent, structural, and unconscious aspects of the document are more reliable indicators of original authorship.

JPlag is not perfect, but has an impressive record, even when deliberate attempts are made to deceive it. The program has also been subject to efforts to optimize it in terms of performance and operation times.

YAP

YAP predates JPlag by some years, and the current version, YAP3, uses algorithms very similar to those found in JPlag. YAP3 is fairly extensible and can be used with English text, as opposed to being limited to program source code. YAP is also effective at dealing with code where modules have been transposed or rearranged. This latter capability could be very useful in dealing with polymorphic malware.

YAP originally used UNIX utilities and was implemented as a UNIX shell script. The second version was rewritten in Perl, in order to employ improved and more flexible comparison algorithms.

As with JPlag, YAP attempts to convert the material under study into abstract symbols before analyzing it for similarities. As noted, this allows the analytical engine to compare material regardless of the source language. YAP does not do a full parsing of the content, but looks at keywords reserved for that language. This approach allows YAP to work with English texts as well as code: A simple parsing can find appropriate keywords in the samples of English, and then generate the symbol strings for analysis. As such, YAP may form the basis for a number of applications requiring the matching of patterns, such as DNA sequencing and the reconstruction of shredded or damaged documents.

Other Approaches

Cogger

Cogger, by Padraig Cunningham of Trinity College Dublin and Alexander N. Mikoyan of Moscow State University, seeks to address the particular problem of determining what keywords or markers in the original text are most likely to be effective when generating abstract symbols for duplication analysis. It uses a case-based reasoning approach that originates in the study of artificial intelligence and neural networks. (Cogger is an Irish slang term for the copying of homework or assignments.)

Finding Duplicated Code

The Software Composition Group of the University of Berne is working with a view to similar ends, but with a different purpose in mind. Programmers developing software often duplicate code, whether their own or others, for a variety of reasons. Copying code is simpler than writing it from scratch; it may be easier to just copy code into a new location, rather than turn an existing section into a procedure; existing code is more likely to work without creating problems; and the programmer, possibly judged on the basis of the number of lines of code produced, may simply be bucking for a raise. Duplicated sections of code in a program not only extend its length and waste space (as well as memory and compilation time), but make maintenance of the program more difficult in the future.

Determining duplications in the code is not an easy task. Tools for finding copied sections exist, but they generally require particular tuning for the exact language, and even version, of the source code under study.

The team in Berne is, therefore, looking at ways to find duplications regardless of the specific language being used. This approach has an obvious application to the plagiarism detection systems that we have seen in the other programs, because it eliminates the initial parsing step that all of them require.

The approach taken is not quite language agnostic. There is a canonicalization stage where all comments and white space are removed. This is relatively simple, requires minimal knowledge of the language, and minimal transformation of the material under study.

A first pass at the code matches individual lines. The matrix of the individual matches found is then visualized with a scatterplot. Using a pattern of dots, the individual matching lines can quickly and easily be used to find large sections of similarity. Patterns of diagonal lines of dots indicate sequences of duplicate lines, and therefore sections of duplicate code, as shown in Figure 4.1.

The plot shown in Figure 4.1 indicates a large section of duplicate code that has not been changed or modified for local use. There are other visual patterns that can indicate different uses of duplicate code. Figure 4.2 shows a pattern where the diagonal has been broken, indicating that a small portion of the code has been amended, possibly with a new variable or datum. An insertion or possibly deletion of code is indicated by the diagonals being broken by a random section, but continuing afterwards, as in Figure 4.3. Rectangular patterns of diagonals, as shown in Figure 4.4, demonstrate repeated reuse of sections of code.

The use of visualization, rather than automated detection and extraction of code, provides two benefits. The first is that the patterns indicating duplications are very striking to the human eye and can therefore be detected quickly, even in very large programming projects. A second advantage involves the duplication of code in unusual ways: These would have to be identified in the original programming for the system, and therefore might be missed. With visualization, unusual patterns stand out and call for further investigation. (A similar use of visualization is made in the cusum technique of forensic linguistics, which will be examined in Chapter 8.)

More information about the project and research can be obtained at http://www.iam.unibe.ch/_scg/.

Source Code Searching and SCRUPLE

As noted, there are multiple ways of performing any given task in programming. In gauging the intent of a program, it would be help-

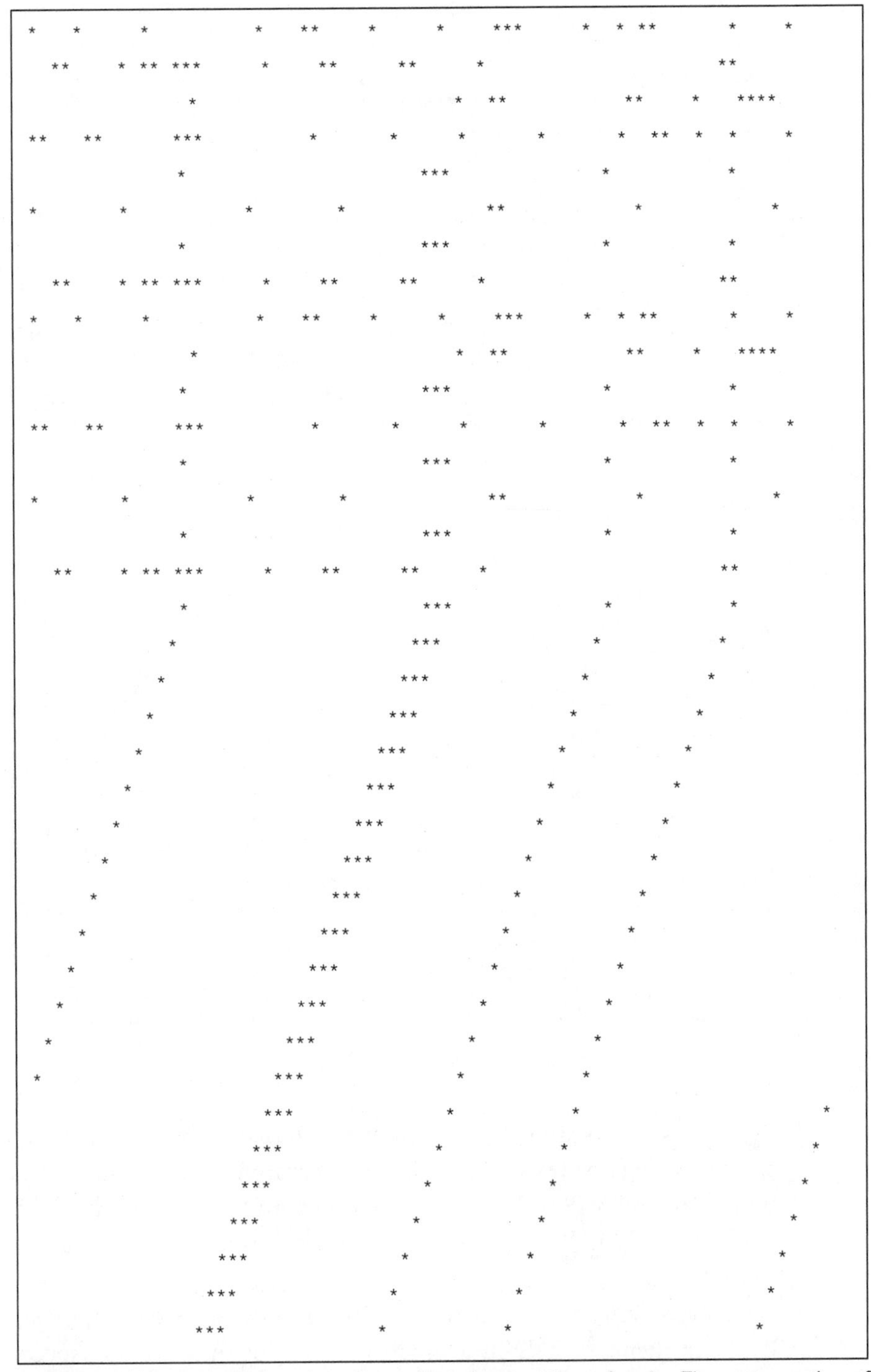

■ **Figure 4.1** *Scatterplot of duplicate lines in a section of code. The upper region of the plot indicates some similarities but no overall duplication, whereas the lower section indicates a large sequence of duplicated code.*

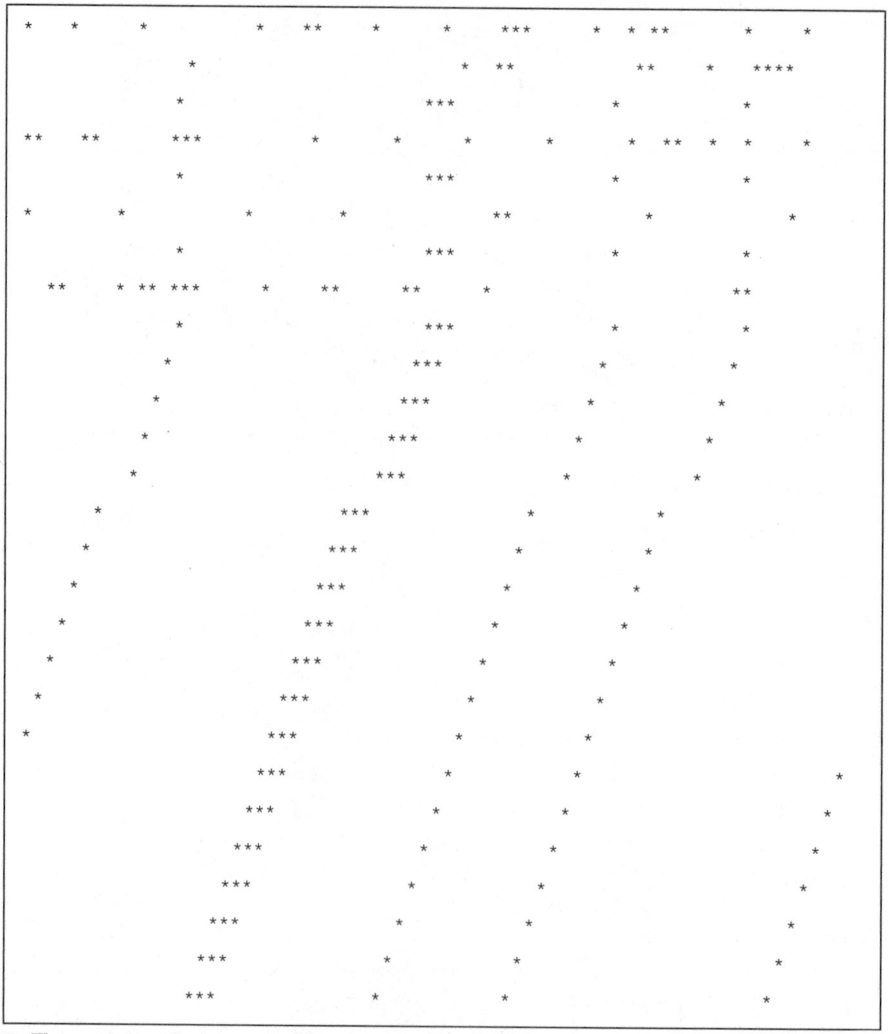

■ **Figure 4.2** *Scatterplot showing a section of duplicate code that has been altered.*

ful if we could search for functional patterns, rather than simply specific strings of text. Maintenance programmers need this capability too, and SCRUPLE, by Paul Santanu and Atul Prakash of the University of Michigan, was written to explore a possible solution to this need.

Pattern languages are a new approach to software development. By using them, SCRUPLE is able to search for operational aspects of the code, as opposed to specific statements or other text elements. Pattern languages can also be used to search for certain styles of

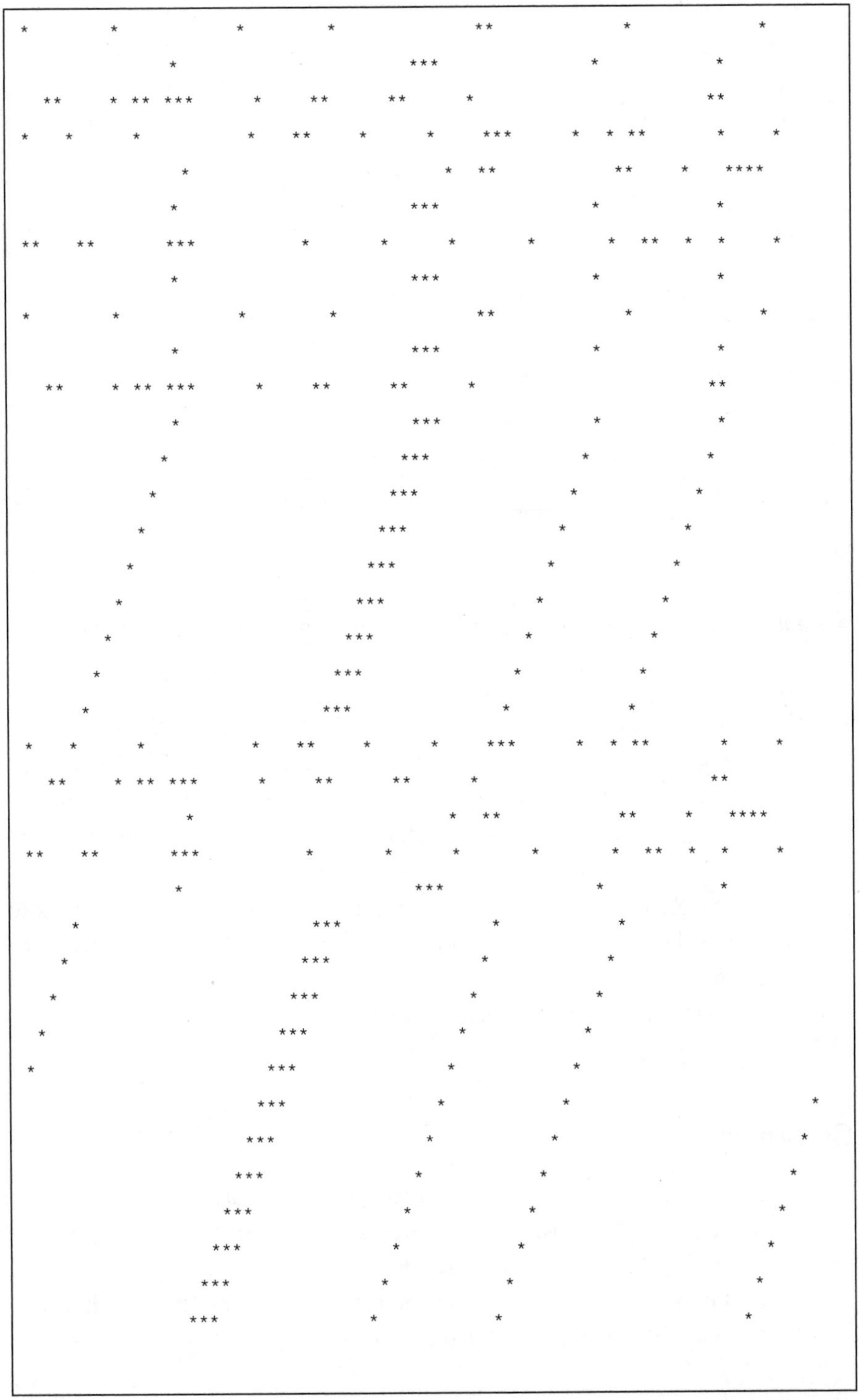

■ **Figure 4.3** *Scatterplot of duplicate code indicating insertion of additional code.*

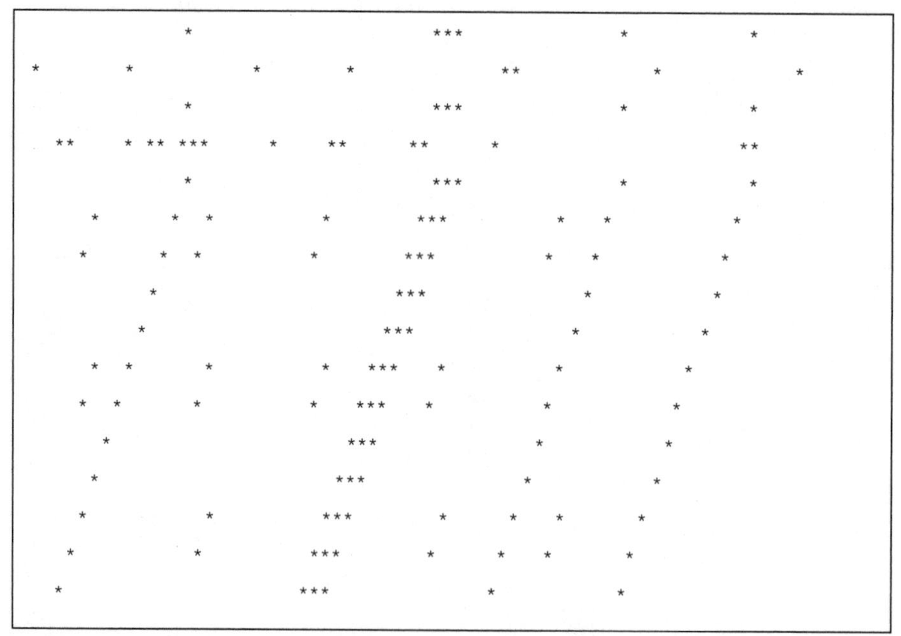

■ **Figure 4.4** *Scatterplot for repeated reuse of code showing rectangular patterns.*

programming, such as bad practices like repetitive code (that should be replaced with calls to a single function) or masses of nested loops (which could be replaced with a case statement). Patterns can also be used to search for specific characteristic errors and bugs, which may be used to identify an author or group.

SCRUPLE has been implemented to deal with both C and PL/AS languages. Pattern libraries for other languages could be created. It may be possible to build patterns for object code as well, so that searches for specific functions could be made directly in binary programs.

Summary

Advanced tools for software forensics do exist, but in very rudimentary form. Most of these utilities are experimental; to operate properly, they require a good deal of tuning and expertise. These preliminary packages indicate the features and capabilities that may be available in the near future.

5

Law and Ethics—Software Forensics in Court

First, because this chapter focuses on legal issues, I probably need to make legal disclaimer-type noises. Therefore, be it known to all that I am not a lawyer, I have never even played one on TV, this is not to be considered legal advice, for legal advice please see qualified legal counsel, void where prohibited by law, no warranty express or implied is made on the fitness of this information for any purpose including the purpose for which it was intended, no added salt, your mileage may vary, this product contains not less than 70 percent recycled opinions, please do not read while operating heavy machinery, have I missed anything?

Legal Systems

In dealing with legal issues, we have an immediate problem in that not only do different countries have distinct laws, but possibly even diverse legal systems.

Those from Britain, the Commonwealth countries, and the United States will be most familiar with the "Common Law" system, based on the presumption of laws that uphold the common good, from an originating charter document and case law precedents laid down over the years. (Common Law is also the system under which a suspected criminal is presumed to be "innocent until proven guilty.") In some of those countries, there are specific laws that would make, for example, malicious software illegal. In Canada, I believe the relevant section of the Criminal Code states that anyone "who, without authorization, modifies data, or causes data to be modified" is guilty of an offense, which would seem to cover it nicely. However, most Common Law systems also have provisions against mischief or vandalism, so malicious software could probably be prosecuted even in the absence of a specific law. (Successful prosecution is quite another matter, the requirements for which we will be dealing with at length.)

Some countries, such as France, have Code Law or Civil Law systems. Under these systems, an activity is not illegal unless there is a specific law against the activity. (In access control terms, everything is permitted unless it is forbidden.) Therefore, under such systems, it may be perfectly legal to write and distribute malicious software (or break into computer systems, or sell pirated copies of copyright protected software) simply because there are no laws prohibiting such activities. The lawmakers haven't caught up with the times. (Blackhats need not think they can get off freely by moving to such countries. If you travel in or to a country where your activity is illegal, you can be prosecuted there. And sometimes they can even extradite you.)

Common Law and Code Law systems are not the only types. There are religious legal systems, and you may also encounter systems based on socialist theories of social and economic structures.

Differences within Common Law

Just to make the situation even more obscure (given lawyers, what else did you expect?), beware of confusing the two different uses of the term "civil." Common Law and Code (Civil) Law are two different types of legal systems. Under the Common Law system, there are criminal cases, tort (or civil) cases, and regulatory cases, governed by different types of law. Criminal cases involve criminal laws and possibly jail time. Civil cases involve some form of tort, or injury (which just has to be a hurt, it doesn't have to be a wound) to someone. Once again, under the Common Law system, even if you can't, for some reason, prosecute someone as a criminal for writing malicious software, if you can prove that what he or she did hurt you in some way (cost you money, lost you something, or even just got people to make fun of you), then you can launch a civil action, or lawsuit, against him or her. You can't put him or her in jail, but you may be able to get some money. (Well, you can get a decision that they owe you something. Collecting actual money may be an entirely different issue, indeed.)

There is an additional point to be made about the difference between civil and criminal cases under Common Law systems, and it is directly relevant to forensic studies. The test of evidence and proof is not the same in the two types of cases. A criminal case must be proven "beyond a reasonable doubt." Civil cases only require that a decision be made on the balance of the probabilities. Thus, evidence for a criminal trial must be presented much more carefully.

Jurisdiction

When dealing with information and telecommunications systems, there may be issues of different jurisdictions, that is, determining who has the right to try the case, and under what system of law. For example, if a Canadian living in Canada started a business writing malware that was created and distributed from a machine in the United States, and that malware affected someone in Britain, it is possible that the person responsible for creating the malicious software could be prosecuted in any or all of the three countries. A greater problem arises when the activity could be legal in one jurisdiction, but is against the law in another.

Jurisdictional issues can be extremely complex, and everyone involved in Internet activities should be aware of the potential problems. Courts are increasingly willing to allow cases to proceed even if the activities of an individual or site are intended for a specific area, and legal in that locale, simply because they are available on the Internet and may be illegal elsewhere.

A similar, but unrelated concept is that of venue, the particular place where the trial is conducted. In many cases, a great deal of effort and maneuvering is used to ensure that a trial proceeds at a specific location, either for the convenience of one side, or to increase the cost and difficulty, in relation to attendance and the production of witnesses, to the other.

Evidence

Simply by including the word "forensic" in the title phrase indicates that we are primarily concerned with evidence. That being the case, we need to know what items the law and the courts consider to be evidence.

Types of Evidence

Many types of evidence can be offered in court. The most common forms of evidence are direct, real, documentary, and demonstrative.

Direct evidence is the normal statement from a witness that we are all familiar with from television dramas. Knowledge is obtained from any of the witness's five senses and is, in itself, proof or disproof of a fact. Direct evidence is called to prove a specific act or occurrence.

Real evidence, also known as associative or physical evidence, is made up of tangible objects. Physical evidence includes such things as tools used in the crime, fruits of the crime (possession of stolen goods, or even the sudden increase of a pile of cash), or possibly perishable evidence that we may be able to reproduce. Often the purpose of the physical evidence is to link the suspect to the scene of the crime. It is this evidence that has material existence and can be presented to the view of the court and jury for consideration.

Documentary evidence is material presented to the court in the form of business records, manuals, printouts, and so forth. Much of the evidence submitted in a computer crime case is documentary, and there are special considerations for this type of material, which we will be examining later.

Finally, demonstrative evidence is evidence used to aid the jury. It may be in the form of a model, experiment, chart, or an illustration offered as proof. It should be noted that to aid the court and the jury, demonstrative evidence is being used more often, especially in the form of simulation and animation. In regard to software forensics, there will be a significant requirement for demonstrative evidence until such time as the courts are willing to accept the process as reliable. Explanations of how software forensics works, how the characteristics are obtained and analyzed, and the reliability of the procedures (including error rates) are all forms of demonstrative evidence.

There are many mathematical algorithms used in software forensic analysis that must either be stipulated or proven to the court to be completely accurate. It is generally more difficult to admit a simulation as evidence because of the substantive nature of the process. Simulations must make a great many assumptions about what is important in the real world, and it is unlikely that the opponent in litigation would allow such assumptions to go unchallenged. Computer animation, on the other hand, is simply a computer-generated sequence, illustrating an expert's opinion. (We will discuss, in more depth, the requirements for expert testimony later in this chapter.) Animation does not predict future events. It merely supports the testimony of an expert witness through the use of demonstrations. An animation of a hard disk spinning, while the read/write heads are reading data, can help the court or jury understand how a disk drive works. There are no mathematical algorithms that must be proven. The animation solely aids the court and jury through visualization. The key to having animation admitted as evidence is in the strength of the expert witness. It is very important to understand the difference between these two types of evidence because it affects the standard of admissibility (discussed in the next section).

Rules of Evidence

The specific rules of evidence may vary between jurisdictions, but some principles are fairly standard. Some are common sense, while others are not quite as obvious, and may direct specific actions that we must take in regard to the collection, preparation, and presentation of exhibits.

First, evidence must be relevant. While the assessments that are done on software may be technically interesting, the court will not want to hear about results unless the data have a bearing on the matter at hand. The fact that a given program uses a particular programming trick is not necessarily evidence. The fact that the trick is used in this program, and only in programs written by one particular author, is evidence. Analysis of software can provide information about what a program does, what it was intended to do, and possibly who wrote it, but generally there will need to be additional data (possibly about the programming environment) to make this information into evidence. On the other hand, a comparative analysis of the code of two programs can probably provide evidence of whether one was copied from another, and which came first.

Another concept in evidence is what is called a "foundation of admissibility." This is somewhat more technical, and deals with the reliability or acceptability of the evidence as something upon which the court can base a decision. It partly relates to specific procedures for what is called the "chain of evidence" or "chain of custody" and also the matter of expert testimony, both of which we will examine shortly.

Any information presented as evidence should have been obtained by legal means. The results of illegal searches or surveillance will not (with some provisos) be accepted in court. Some recent laws may have interesting ramifications in this regard. The United States Digital Millennium Copyright Act (DMCA) states that it is illegal to break any technology intended to protect copyright material. Programmers are held to have copyright on the programs that they produce. Therefore, if the programmer even does a simple self-encryption on the program code, it may be illegal to do a decryption to analyze the software.

Hearsay and Business Documents

An additional factor in the admissibility of evidence is the concept of hearsay. As a witness in court, you may be asked to say what you did, or directly witnessed. Except in very unusual circumstances, you will not be asked, and will not be allowed to say, what someone else told you that he or she did or saw. This "secondhand" testimony is called hearsay and is pretty much automatically suspect. If the court is to accept evidence other than directly from the source, there has to be corroborating testimony.

Business documents are all considered to be, in some sense, hearsay. This is because they are all, in a way, information about a transaction, rather than direct evidence of a transaction. This is particularly true in relation to electronic data. When presenting printouts or other representations of digital information, there must be testimony about how the information is stored and handled, whether there are regular procedures, whether there was any kind of departure from regular procedures in this particular instance, what protections are in place to ensure the integrity of the data, who has access and the ability to change the data, and so forth.

Digital Frailty and the Chain of Evidence

Digital information is extremely fragile. A change of a single bit in a megabyte file can radically alter implications, meaning, or outcomes. While operating systems generally track changes to some extent, even if only the last time a file was changed, it is quite possible to alter data without leaving any trace that a modification has been made, or by whom. Therefore, it is vitally important, in terms of admissibility, to be able to prove that the analyst has not made any changes to the system while studying it. This is so important that the G-8 nations commissioned an initiative for a set of proposed international principles for computer evidence. The standards state that actions taken in seizing digital evidence should not change the evidence, that any person accessing the evidence should be forensically competent, and that all actions in regard to the evidence should be fully documented.

Of course, not every single-bit change to a massive file has any significance. In fact, there are a number of situations in which identical files are stored in different ways, without any meaningful change having taken place at all. For example, text files in DOS and Windows systems are stored with two characters (a carriage return and a line feed) at the end of every line. UNIX file systems store text files with only a single character indicating the end of the line. Therefore, for identical text files stored on different file systems, the hash value, digest, or digital signature would detect a difference (because they are designed to indicate a variation even in a single bit) and declare the files to be different. Thus, computer forensics must sometimes deal with the issue of "canonicalization," which ensures that the digital material being marked to determine changes is limited to those sections where changes are relevant. Obviously, the concept of canonicalization itself needs to be presented carefully in court, if relevant, to ensure that the significance and reliability of the procedure is understood.

An important concept in the presentation of any physical evidence is the "chain of custody." The court will need to know who had access to the evidence and what they did to it. The court will also need to know that nobody other than the people listed could have had access to the material. If there is any possibility that the evidence might have been tampered with, even if there is no specific reason to suspect that something was done to the items in question, then there is a reasonable doubt that the evidence actually does support what we assert.

In the case of digital and electronic data, the identification and preservation of the information is absolutely vital. Digital data are fragile. Data can be easily modified, and, once modified, there is almost no way to determine that a change was made, when the change was made, and who made it. Therefore, the establishment of an ironclad chain of custody is crucial.

The chain of custody for software forensics is going to use procedures very similar to those for other forms of electronic evidence. There will have to be identification and preservation of the original system. Either the computer itself or the electronic media will need to be labeled and secured. Access to these original items must be restricted. (In some special circumstances, of course, the material to be studied will be available only in the memory of the computer, and therefore, the original physical representation cannot be preserved. In that case, the procedures involved in recovering the material, and maintaining it thereafter, must be documented.) Printouts and/or message digest calculations should be created as soon as possible to demonstrate that the data studied or presented are the same as the original. The original material should not be studied because the tools used to study the files can also be used to modify them. Therefore, copies should be made and verified (possibly through the use of digest calculations) to be identical to the originals.

Providing Expert Testimony

For the foreseeable future, software forensics will be an arcane art that is presented in court, if at all, by an expert. Therefore, the rules governing expert witnesses are also germane to our discussion.

Despite the aphorism (attributed to Mark Twain) that an expert is simply some guy from out of town, it is not necessarily possible to walk into court and simply state that you are an expert in a field. The court will decide (or the other side will challenge) whether you have sufficient education and training, experience, or skill for your conclusions, and a "reasonable basis" for the conclusions you reach. If you aren't an expert, you can present the uninterpreted results of your analysis. However, masses of unexplained data are likely to be judged irrelevant to the case. If you do obtain the status of an expert witness, then you are also able to present your opinion as to what the results mean.

Assessment of expertise in cases involving information technology has been, and is currently difficult. Unlike other professions such as medicine, civil engineering, or accounting, there is no organized, licensed professional college, organization, institute, or other body involved in the regulation, censure, and discrimination of specific expertise. (Or, at least, no universally recognized one. As the old joke has it, the nice thing about computer standards is that there are so many of them.) Given the novelty of the field of software forensics, the requirements for education and training are likely to be even more problematic. After all, there is no certificate, diploma, or degree program in software forensics. Some individuals are working on various aspects of the field, but academic research in isolated areas is unlikely to be accepted as sufficient training: at least not in and of itself. The expert need not have complete knowledge about the field in question, need not be certain, and need not be unbiased: The expert must only be able to aid the jury in resolving a relevant issue. While the level of expertise may affect the weight to be accorded the expert's opinion, it does not affect admissibility. It has been held that the court cannot exclude testimony simply because the trial court does not deem the proposed expert to be the best qualified or because the proposed expert does not have the specialization the court considers most appropriate.

Again, experience will be difficult to assess in the immediate future. Work as a low-level programmer analyzing and debugging code is likely to be a help, but not necessarily sufficient. Academic research on plagiarism and the similarity of code between programs could be relevant. Virus researchers, used to finding identifiable patterns in programs (as well as specific functions), could have an advantage. However, the experience of a specific software forensic practitioner will probably have to be matched with the particular case being pursued.

Skill is an interesting concept. It is not necessarily simply related to training or experience. A peer group may very well adjudge an individual as being particularly skilled, even though there is little or no difference in training or experience between the skilled person and his or her colleagues. Therefore, we may be faced with the situation of bringing testimony, of one type or another, attesting to the skill of the expert being brought before the court. A demonstrated success rate may be said to be a measure of skill, but even this will be a rather subjective criterion because definitions of "success" will vary.

Of the different factors that make up the assessment as to whether a witness is expert or not, probably one that deserves a lot of attention is that of reasonable basis. It is not enough that we, as experts, can see a conclusion or finding. An expert's opinion must be reliable; that is, based on valid reasoning and reliable methodology, as opposed to subjective belief or unsupported speculation. If an expert opinion is based on speculation or conjecture, it may be stricken. Expert testimony cannot be based on pure speculation, rather than reasonable inference. We must be able to explain the reasoning leading to that conclusion to at least a judge, and possibly a jury, who probably do not have any technical background. Software forensic evidence is likely to rely on statistical methods. Presentation of the evidence, therefore, will need to be accompanied by clear explanations not only of how the characteristics were determined, but also of why they are important. The court must make an assessment as to whether the reasoning or methodology underlying the testimony is scientifically valid. To do so, the court should consider whether a method consists of a testable hypothesis, whether the method has been subject to peer review, the known or potential rate of error, the existence and maintenance of standards controlling the technique's operation, whether the method is generally accepted, the relationship of the technique to methods that have been established to be reliable, the qualifications of the expert witness testifying based on the methodology, and the nonjudicial uses of the method. All of these factors should be addressed in presenting expert testimony in court.

In the United States, there is case law, known as the Daubert decision, that states judges must decide not only whether a witness is truly expert, but on what the expert may testify about or to. This restriction of testimony on the part of the judge has become known as the "gatekeeper function." In American courts, an expert witness may find that restrictions are placed on the testimony that can be delivered. The court, even after having decided that a witness is an expert, may limit the areas of expertise that are acceptable.

An expert witness may have a number of roles in regard to a case. First is the consulting expert, who provides advice and information to counsel well in advance of the trial itself. At the time of the trial, you may be asked to act as the court's expert, giving supposedly unbiased explanations to the court (usually the judge) about specialized technical matters bearing on the case. Then there is the testifying expert, which is the role most people would think of in regard to expert witnesses. The testifying expert explains findings, discoveries, and their implications. However, the discoveries may be such that they could only be seen or made by the expert. In that case, the expert witness also functions as a witness to fact. It is important to ensure that the expert knows, and does not confuse, which of these roles he or she is being asked to undertake. The expert may, in fact, be asked to fulfill more than one of these roles in the course of proceedings, and in such a situation, maintaining the division between roles is vital. The function the professional is being asked to undertake will shape the work to be done and will possibly restrict the testimony that can be given.

In normal testimony, witnesses are only allowed to tell what they saw, heard, or in some way know to be a fact. Witnesses are not permitted to infer or extrapolate beyond direct observation. Experts are allowed to draw conclusions and even present opinions. This extension of the testimony of the professional is not unrestricted and will be more open to challenge by the opposing side.

One very common source of tension in obtaining and presenting technical testimony is the significant difference in mindset between the technical and legal worlds. Computer work generally involves finding an answer to a problem: If the code works, background study and documented analysis is generally irrelevant. The legal profession, on the other hand, depends absolutely on advance preparation, and an answer is almost useless unless the reasoning, background, and process is not only chronicled, but properly and legally obtained. Both the legal counsel and the technical expert will have to work to see the other's point of view. Lawyers, used to carting around trunks full of papers for even the simplest case, will have to be prepared for the assumption that "it works" is sufficient proof. Techies will need to overcome their deep-seated aversion to any and all forms of documentation, and the automatic assumption that all users are incapable of understanding technical concepts.

Ethics

The topic of ethics is a very difficult one, and a complete discussion is beyond the scope of this book. Discussions of ethics, particularly in regard to information technology, tend to fail to provide useful direction. This deficiency is due in large measure to a breakdown in communication: The parties to such debates seldom agree even on the most basic definition of what ethics actually are. Some see ethics as fundamental guiding principles, others assume that they are legalistic codes of conduct. Few can even agree on whether ethical standards are absolute or relative. I will not be pursuing this deliberation, beyond the brief review of the variant hacker ethics, from Chapter 2. However, some related topics are more directly relevant to software forensics.

Disclosure

The concept of disclosure—how much information to provide about a potential security weakness—is one that has become a major issue in security circles in recent years. In the early days of computer security work, a majority of practitioners came from a military or related background, with a heavy emphasis on confidentiality. Thus, it was automatically assumed that limiting information was a good thing. Because nobody knew much about computers, keeping the details quiet did provide a limitation on the number of people who could even attempt to penetrate systems. Latterly, this position has become known as security by obscurity or SBO.

Obscurity has been found wanting. In the same way that writers cannot edit their own text (trust me on this), security planners and administrators are often blind to the faults in their own systems.

Without some outside analysis of a setup or product, flaws can go undetected for years. In such a situation, the first time the existence of a problem becomes apparent is when someone uses it. The attackers and blackhats will find weaknesses, even if nobody talks about them. The only people that obscurity keeps in the dark are those charged with protecting the systems.

Some advocate full disclosure. This means that anyone who discovers a security loophole should immediately publish all details in full, including instructions on how to use the vulnerability. Because full disclosure requires that everyone be alert to every security warning that comes along, no matter how minor, most people are more comfortable with some level of partial disclosure. Partial disclosure usually involves limits, such as informing the publisher of a product before alerting the general public to give the vendor time to come up with some way to fix the problem. Sometimes the information given to the public may be restricted to the existence of the issue, plus suggestions about safety measures. Partial disclosure is not standardized in any way, and the definition of what it entails may vary from person to person.

The virus research community has frequently been accused of practicing a form of security by obscurity. Legitimate researchers refuse to distribute virus code unless they know that the person making the request is qualified, and that the requester abides by the same code of conduct and will not give out copies of the software. This position may seem untenable in these days of email viruses, when virus code may be obtained almost as easily as spam messages. The community, however, believes that establishing a flexible line on distribution would be too complex and would inevitably lead to increased distribution and therefore chooses to err on the side of caution. Keep this in mind when contacting virus researchers and asking for assistance with malicious code.

Blackhat Motivations as a Defense

It may be important to examine the commonly presented justifications for blackhat activity as arguments relating to the activity of software forensics. Some of these contentions may be used as a defense in court. Regardless of how we may consider them from an ethical standpoint, we may need to be prepared to defend against them as legal debates.

One of the most frequently attempted justifications of blackhat activity of all kinds is that it is protected under the concept of freedom of speech. The free speech defense may be an extremely strong one, particularly in nations under Common Law systems, where some form of freedom of expression is frequently a constitutional right. So far, the courts still appear to be divided on the issue of whether computer code counts as "speech." A more reliable approach to dealing with the free speech argument is to examine the restrictions on freedom of speech, usually restricting the right to create harm with speech.

Many individuals who practice system violation activity explain themselves on the basis that they are following in the detective footsteps of the old-time hackers, who explored and discovered the capabilities of early computing devices. In many cases, there is honest disagreement between individuals in regard to the legality of attempts to break security systems for the purpose of strengthening those same systems. Some individuals who seem to have sincerely wanted to publicize security weaknesses, or offer their services as consultants, have found themselves facing criminal charges. There are two points to make in this regard: One is to ensure that the activity you are examining is malicious in intent. The other is to be very sure that, when investigating a system, you have permission to do so.

Summary

In today's interconnected computing environment, you may be faced with a variety of legal systems. While you cannot know all the relevant laws around the globe, a general overview of the types of legal systems you may encounter can be helpful.

As the title of the book implies, the software forensic investigator is collecting, analyzing, and presenting evidence. Note that there are very definite rules in regard to the collection and treatment of evidence. In particular, note the chain of custody concept that is so vital in the traditional data recovery side of computer forensics.

As a specialist in software forensics, you may be required to act as an expert witness, in one or more roles. Be aware of the special demands and responsibilities of that function.

A number of issues will bear on your activities in the legal arena. While law and ethics are not identical, it is probably necessary to examine all activities from an ethical perspective while the legislation and precedents relating to software are being worked out.

6

Computer Virus and Malware Concepts and Background

Computer viruses are unique among the many security problems in that if someone else is infected, it increases the risk to you. Malware also seems to be surrounded by myths and misunderstandings. The purpose of this chapter is not to explain virus concepts for their own sake. The intent is to present the structure normally found in viruses and malware, as a guide to analysis of what may be the largest single class of software objects subject to forensic analysis. While analysis of malware is not the only use for software forensics, evaluation of the intent of a program uses most of the techniques involved in the field as a whole.

At the same time, note that computer virus research gave rise to what is known as forensic programming, one of the earliest forms of software forensics. A study of the history and investigation of computer viruses, therefore, can be very helpful as background to the practice of program analysis.

History of Computer Viruses and Worms

Many claims have been made for the existence of viruses prior to the 1980s, but so far, these claims have not been accompanied by proof. The Core Wars programming contests did involve self-replicating code, but usually within a structured and artificial environment. Discussions around this point may have academic interest, but are of little consequence in regard to forensic programming because any self-replicating programs would have been quite readily identified as to author and intent.

The general perception of computer viruses, even among security professionals, has concentrated on their recent existence in personal computers, and particularly Wintel type systems. This is despite the fact that Fred Cohen's seminal academic work took place on mainframe and minicomputers in the mid-1980s. The first email virus was spread on mainframes in 1987. The first virus hoax message (then termed a metavirus) was proposed in 1988. Even so, virus and malware research have been neglected, possibly because malware does not fit easily into the traditional access control security models.

Dr. Cohen's research and writings are quite important in regard to forensic analysis. One of the major points of his study dealt with the question of decidability: the issue of whether we can determine that a program is viral. Cohen's analysis clearly proved that we never can know, in advance, whether a program is a virus. Any algorithm we use will be subject to flaws: either it will have a potentially infinite number of false positives (identifying a program as a virus when it is not), or an infinite number of false negatives (failing to identify a program that is a virus), or both. In practical terms, this means that there are no "easy answers" in forensic analysis, particularly in regard to the intent of a program. There are no specific markers that we can use to determine if a program is viral or malicious. Not only does this make the analytical work harder, but it also means that presentation of the evidence in court will be subject to explanation and interpretation.

During the early 1990s, virus writers began experimenting with various functions intended to defeat detection. (Some forms had seen earlier limited trials.) Among these was polymorphism, to change form in order to defeat scanners, and stealth, to attempt to confound any type of detection. None of these virus technologies had any significant impact. Most viruses using these "advanced" technologies were easier to detect because of a necessary increase in program size. While not effective as a means of avoiding detection, the widespread use of packing, compression, and other forms of simple encryption makes the forensic process more time-consuming.

Although demonstration programs had been created earlier, the mid-1990s saw the introduction of macro and script viruses in the wild. These were initially confined to word processing files, particularly files associated with the Microsoft Office Suite. However, the inclusion of programming capabilities eventually led to script viruses in many objects that would normally be considered to contain data only, such as Excel spreadsheets, PowerPoint presentation files, and email messages. This fact led to greatly increased demands for computer resources among antiviral systems because many more objects had to be tested, and Windows object linking and embedding (OLE) format data files presented substantial complexity to scanners. Macroviruses also increase new variant forms very quickly because the virus carries its own source code, and anyone who obtains a copy can generally modify it and create a new member of the virus family. From a forensic perspective, script and macromalware are fairly simple objects for analysis. Because the source code is available, the techniques of code and text examination are readily applicable.

Email viruses became the major new form in the late 1990s and early 2000s. These viruses may use macrocapabilities, scripting, or executable attachments to create email messages or attachments sent out to email addresses harvested from the infected machine. Email viruses spread with extreme rapidity, distributing themselves worldwide in a matter of hours. Some versions create so many copies of themselves that corporate and even service provider mail servers are flooded and cease to function. Email viruses are very visible, and so tend to be identified within a short space of time, but many are macros or scripts, and so generate many variants. On the surface, it would seem that network forensics could be used to trace the authors of email viruses. In some cases, this can and has been done. However, because email viruses send and resend themselves from every machine infected, it may take significant effort and resources to trace back to the original release. Also note that many instances of email viruses have included references to Web pages, email addresses, or other labels intended for updating or tracking of the virus, but which can also be used to identify the author.

Various types of malware have been listed in Chapter 2. Given the strong integration of the Microsoft Windows operating system with its Internet Explorer browser, Outlook Mailer, Office Suite, and system scripting, recent viruses have started to blur the normal distinctions. A document sent as an email file attachment can make a call to a Web site that starts active content, which installs a remote access tool acting as a portal for the client portion of a distributed denial of service (DoS) network. Indeed, not only are viruses beginning to show characteristics that are similar to each other, but functions from completely different types of malware are beginning to be found together in the same programs, leading to a type of malware convergence. This will, obviously, complicate software forensic analysis, but it is still important to keep the types and functions of specific malware examples in mind.

Recently, many security specialists have stated that the virus threat is reducing because despite the total number of virus infections being seen, the prevalent viruses are now almost universally email viruses, and therefore, constitute a single threat with a single fix. This ignores the fact that while almost all major viruses now use email as a distribution and reproduction mechanism, there are a great many variations in the way email is used. For example, many viruses use Microsoft's Outlook Mailer to spread, and reproduction can be prevented simply by removing Outlook from the system. However, other viruses may make direct calls to the mail application programming interface (MAPI), which is used by a number of mail user programs, while others carry the code for mail server functions within their own body. A number of email viruses distribute themselves to email addresses found in the Microsoft Outlook address book files, while others may harvest addresses from anywhere on the computer hard drive, or may actually take control of the Internet network connection and collect contact data from any source viewed online. Again, it will be very important in forensic analysis to note any subtle distinctions in the way certain operations are invoked, so as to be able to distinguish between authors who probably have preferred methods.

While the idea is controversial, no less an authority than Fred Cohen has championed the idea that viral programs can be used to great beneficial effect. An application using a viral form can improve performance in the same way that computer hardware benefits from parallel processors. It is, however, unlikely that viral programs can operate effectively and usefully in the current computer environment without substantial protective measures being built into them. A number of virus and worm programs have been written with the obvious intent of proving that viruses could carry a useful payload, and some have even had a payload that could be said to enhance security. Unfortunately, all such viruses have created serious problems themselves. As noted previously, with respect to the question of decidability, the fact that technologies used in malware can also be used in legitimate software will make presentation of forensic evidence very complicated.

Malware Definition and Structure

Malware, a term formed from the contraction of the phrase *malicious software* is used to describe software or programs intentionally designed to include functions for penetrating a system, breaking security policies, or carrying malicious or damaging payloads. As noted, specific types of malware have been listed and described in Chapter 2.

It is sometimes difficult to make a hard and fast distinction between malware and bugs. For example, if a programmer left a buffer overflow in a system and it creates a loophole that can be used as a backdoor or a maintenance hook, did he or she do it deliberately? This question cannot be answered technically, although we might be able to guess at it, given the relative ease of use of a given vulnerability.

Note that malware is not just a collection of utilities for the attacker. Once launched, malware can continue an attack without reference to the author or user, and in some cases, will expand the attack to other systems. There is a qualitative difference between malware and the attack tools, kits, or scripts that have to be under an attacker's control, and which are not considered to fall within the definition of malware.

Malware can attack and destroy system integrity in a number of ways. Viruses are often defined as needing to attach to programs (or to objects considered to be programmable) and so must, in some way, compromise the integrity of applications. A number of viruses attach themselves to the system in ways that either keep them resident or invoke them each time the machine starts and so compromise the computer or overall network, even if individual applications are not touched. Remote access trojans/tools (RATs) (basically remotely installed backdoors) are designed to allow a remote user or attacker to completely control a system, regardless of local security controls or policies.

(The fact that viruses modify programs is seen as evidence that viruses inherently compromise systems, and therefore, the concept of a "good" or even benign virus is a contradiction in terms.)

Many viruses or other forms of malware contain payloads (data diddlers) that may either erase data files or interfere with application data over time in such a way that data integrity is compromised, and data may become completely useless.

There is also an attack on a type of integrity that is very important in considering malware. As with attacks where the intruder takes control of your system and uses your computer to explore or attack further systems, to hide his or her own identity, malware (viruses and distributed denial of service [DDoS] zombies in particular) is designed to use your system as a platform to continue further attacks, even without the intervention of the original author or attacker. This can create problems within domains and intranets where equivalent systems "trust" each other, and can also create "bad will" when those you do business with find out that your system is sending viruses or probes to theirs.

As noted, malware can compromise programs and data to the point where they are no longer available. In addition, malware generally uses the resources of the system it has attacked, and can, in extreme cases, exhaust central processing unit (CPU) cycles, available processes (process numbers, tables, etc.), memory, communications links and bandwidth, open ports, disk space, mail queues, and so forth. Sometimes this can be a direct DoS attack, and sometimes it is a side effect of the activity of the malware.

Malware programs such as backdoors and RATs are intended to make intrusion and penetration easier. Viruses such as Melissa and SirCam send data files from your system to others (in these particular cases, seemingly as a side effect of the process of reproduction and spread). Malware can be written to do directed searches and send confidential data to specific parties, and can also be used to open covert channels of other types.

The fact that you are infected with viruses, or compromised by other types of malware, can become quite evident to others. This provides indirect evidence of your level of security or lack thereof, and may also create seriously bad publicity.

A computer virus is a program written with functions and intent to copy and disperse itself without the knowledge and cooperation of the owner or user of the computer. A final definition has not yet been agreed on by all researchers. A common definition is, "a program which modifies other programs to contain a possibly altered version of itself." This definition is generally attributed to Fred Cohen from his seminal research in the mid-1980s, although Dr. Cohen's actual definition is in mathematical form. Another possible definition is an entity that uses the resources of the host (system or computer) to reproduce itself and spread, without informed operator action.

Cohen is generally held to have defined the term "computer virus" in his thesis (published in 1984). (The suggestion for the use of the term "virus" is credited to Len Adleman, his seminar advisor.) However, his original definition covers only those sections of code that, when active, attach themselves to other programs. This, however, neglects many of the programs that have been most successful "in the wild." Many researchers still insist on Cohen's definition and use other terms such as "worm" and "bacterium" for those viral programs that do not attack programs. Currently, viruses are generally held to attach themselves to some object, although the object may be a program, disk, document, email message, computer system, or other information entity.

Many people have the impression that anything that goes wrong with a computer is caused by a virus. From hardware failures to errors in use, everything is blamed on a virus. A virus is not just any damaging condition. Similarly, it is now popularly believed that any program that may do damage to your data or your access to computing resources is a virus. Viral programs are not simply programs that do damage. Indeed, viral programs are not always damaging, at least not in the sense of being deliberately designed to erase data or disrupt operations. Most viral programs seem to have been designed to be a kind of electronic graffiti—intended to make the writer's mark in the world, if not his or her name. In some cases, a name is displayed; on occasion an address, phone number, company name, or political party.

The point to make in regard to this discussion of the definition of a computer virus is that clear definitions of the characteristics of a specific type of software will be essential for accurate analysis of a software object under study. To the layperson, the difference between a reproductive and a nonreproductive piece of malware may be meaningless, and he or she may refer to both as viruses. The forensic analyst cannot afford that kind of laxity, and must make more rigorous distinctions, not the least because the characteristics of specific operations make up the signatures under study. The software investigator must go even further, and determine whether the reproductive program requires the intervention or action of the user (a virus) or can use system functions without the involvement of the user (a worm).

Malicious software has six basic elements, although not all may be present in each specific program. Insertion is the means used to become resident in the target system. Avoidance, otherwise known as stealth, is any of a number of methods used to evade detection. Eradication is the means by which the malware removes traces of itself following a trigger. Propagation or replication is considered the province of viruses and worms only, and is the function that defines a program as a virus. The trigger is the event or condition that initiates a payload. The payload is the additional coding, usually with negative consequences, carried by the program.

From this point, we will turn to an outline of the structures and components that comprise the different types of malware. While not directly visible in code, these formations can be used as abstract signatures for determining different types of malware. We will restrict this discussion to those types of programs that are fairly clearly malicious. Backdoors, after all, may have started life as debug modes or other necessary maintenance hooks, and there is a good deal of controversy over the legitimacy of adware and spyware. These latter types of programs occupy gray areas on the software spectrum and provide no clear outlines to assist us in our analysis of the intent of code.

Virus Structure

A virus is defined by its ability to reproduce and spread. A virus is not just anything that goes wrong with a computer, and virus is not simply another name for malware. For example, trojan horse programs and logic bombs (defined elsewhere) do not reproduce themselves. A worm, which is sometimes seen as a specialized type of virus, is currently distinguished from a virus because a virus generally requires an action on the part of the user to trigger or aid reproduction and spread.

The action on the part of the user is generally a common computer operation function, and the user generally does not realize the danger of the action, or the fact that he or she is assisting the virus.

The only requirement that defines a program as a virus is that it reproduces itself. There is no necessity that the virus carries a payload, although a number of viruses do. In many cases (in most cases of "successful" viruses), the payload is limited to some kind of message. A deliberately damaging payload, such as erasure of the disk or system files, usually restricts the ability of the virus to spread because the virus uses the resources of the host system to reproduce itself. In some cases, a virus may carry a logic bomb or time bomb that triggers a damaging payload on a certain date or under a specific, often delayed condition.

In considering computer viruses, three structural parts are deemed important: the replication or infection mechanism, the trigger, and the payload.

Infection Mechanism

The first and only necessary part of the structure is the infection mechanism. This is the code that allows the virus to reproduce, and thus to be a virus. The infection mechanism itself has a number of parts to it.

The first function is to search for, or detect, an appropriate object to infect. The search may be active, as in the case of some file infectors that take directory listings to find appropriate programs of appropriate sizes, or it may be passive, in the case of macroviruses that infect every document as it is saved. There may be some additional decisions taken once an object is found. Some viruses may try to actually slow the rate of infection to avoid detection. Most will check to see if the object has already been infected.

The next action will be the infection itself. This may entail the writing of a new section of code to the boot sector, the addition of code to a program file, the addition of macrocode to the Microsoft Word NORMAL.DOT file, the sending of a file attachment to harvested email addresses, or a number of other operations. There are additional subfunctions at this step as well, such as the movement of the original boot sector to a new location, or the addition of jump codes in an infected program file to point to the virus code. There may also be changes to system files, to try and ensure that the virus will be run every time the computer is turned on. This can be considered the insertion portion of the virus.

At the time of infection, a number of steps may be taken to try and keep the virus safe from detection. The original file creation date may be conserved and used to reset the directory listing to avoid a change in date. The virus may have its form changed, in some kind of polymorphism. The active portion of the virus may take charge of certain system interrupts, in order to make false reports when someone tries to look for a change to the system. There may also be certain prompts or alerts generated, in an attempt to make any odd behavior noticed by the user appear to be part of a normal or at least innocent computer error. We will examine detection and antidetection technologies later in this chapter.

Trigger

The second major component of a virus is the payload trigger. The virus may look for a certain number of infections, a certain date and/or time, a certain piece of text, or simply blow up the first time it is used. As noted, a virus does not actually need to have either a trigger or a payload.

Payload

If a virus does have a trigger, then it usually also has a payload. The payload can be almost anything, from a simple one-time message, to a complicated display, to reformatting of the hard disk. However, the bigger the payload, the more likely it is that the virus will get noticed. Therefore, while you may have seen lists of payload symptoms to watch for—such as text messages, ambulances running across the screen, letters falling down, and the like—checking for these payloads isn't a very good way to keep free of viruses. The successful ones keep quiet. However, a virus carrying a very destructive payload will also eradicate itself when it wipes out its target.

Worm Structure

A worm reproduces and spreads, like a virus, but unlike other forms of malware. Worms are distinct from viruses, though they may have similar results. Most simply, a worm may be thought of as a virus with the capacity to propagate independently of user action. In other words, they don't rely on the transfer of data between systems for propagation that is usually human-initiated, but instead spread across networks of their own accord, primarily by exploiting known vulnerabilities in common software.

Originally, the distinction was made that worms used networks and communications links to spread, and that a worm, unlike a virus, did not directly attach to an executable file. In early research into computer viruses, the terms worm and virus tended to be used synonymously because it was believed that the technical distinction was unimportant to most users. The technical origin of the term worm program, used by Shoch and Hupp in a 1982 paper, matched that of modern distributed processing experiments: a program with "segments" working on different computers, all communicating over a network.

The structure of a worm, therefore, is essentially identical to that of a virus, with an infection component (with associated search, infection, and antidetection mechanisms), and a possible trigger and payload. The only difference is that the invocation or infection phases do not depend on user intervention, but can be accomplished directly by the worm.

Trojan Structure

A trojan, or trojan horse, is a program that pretends to do one thing while performing another unwanted action. The extent of the pretense may vary greatly. Many of the early PC trojans relied merely on the filename and a description on a bulletin board. "Login" trojans, popular among university student mainframe users, mimicked the screen display and the prompts of the normal login program and could, in fact, pass the username and password along to the valid login program at the same time they stole the user data. Some trojans may contain actual code that does what it is supposed to do, while performing additional nasty acts that it does not tell you about.

Some data security writers consider that a virus is simply a specific example of the class of trojan horse programs. There is some validity to this usage because a virus is an unknown quantity that is hidden and transmitted along with a legitimate disk or program, and any program can be turned into a trojan by infecting it with a virus. However, the term virus more properly refers to the added, infectious code, rather than the virus/target combination. Therefore, the term trojan refers to a deliberately misleading or modified program that does not reproduce itself.

A major aspect of trojan design is the social engineering, that is, the fraudulent or deceptive component. Trojan programs are advertised (in some sense) as having a positive component. Social engineering really is nothing more than a fancy name for the type of fraud and confidence games that have existed since snakes started giving away apples. Security types tend to prefer a more academic sounding definition, such as the use of nontechnical means to circumvent security policies and procedures. Social engineering can range from simple lying (such as a false description of the function of a file), to bullying and intimidation (to pressure a low-level employee into disclosing information), to association with a trusted source (such as the username from an infected machine).

The structure of a trojan is, therefore, extremely simple: There is a deception component that gains insertion and a negative payload. Unfortunately for the forensic analyst, software examination alone cannot always identify the deceptive component. The fraud may be perpetrated by a filename or the description on an email subject line. As this book is being written, a rather depressing development in the malware world is the use of small, legitimate, and otherwise helpful utilities, put together in packages intended to create a negative outcome for the recipient.

Pranks are very much a part of the computer culture—so much so that you can now buy commercially produced joke packages that allow you to perform "Stupid Mac (or PC, or Windows) Tricks." Numerous pranks are available as shareware. Some make the computer appear to insult the user; some use sound effects or voices; some use special visual effects. A fairly common thread running through most pranks is that the computer is, in some way, nonfunctional. Many pretend to have detected some kind of fault in the computer (and some pretend to rectify such faults, of course making things worse). One entry in the virus field is PARASCAN, the paranoid scanner. It pretends to find large numbers of infected files, although it doesn't actually check for any infections.

Generally speaking, pranks that create some kind of announcement are not malware: Viruses that generate a screen or audio display are actually quite rare. The distinction between jokes and trojans is harder to make, but pranks are intended for amusement. Joke programs may, of course, result in a DoS if people find the prank message frightening. If you wish to consider a prank as a type of trojan program, the structures are basically identical.

One specific type of joke is the "Easter egg," a function hidden in a program, and generally accessible only by some arcane sequence of commands. These may be seen as harmless, but note that they do consume resources, even if only disk space, and also make the task of ensuring program integrity very much more difficult. An Easter egg is structured slightly differently from a trojan, and is constructed similarly to a logic bomb.

Logic Bomb Structure

Logic bombs are software modules set up to run in a quiescent state, but to monitor for a specific condition or set of conditions, and to activate their payload under those conditions. A logic bomb is generally implanted in or coded as part of an application under development or maintenance. Unlike a RAT or trojan, it is difficult to implant a logic bomb after the fact. There are numerous examples of this type of activity, usually based on actions taken by a programmer to deprive a company of needed resources if employment were to be terminated.

The structure of a logic bomb is fairly simple, consisting of a component to check for a trigger condition and a payload.

A trojan or a virus may contain a logic bomb as part of its payload. A logic bomb involves no reproductive or social engineering components.

Remote Access Trojan (RAT) Structure

RATs are programs designed to be introduced, usually remotely, after systems are installed and working (and not in development, as is the case with logic bombs and backdoors). Their authors would generally like to have the programs referred to as remote administration tools to convey a sense of legitimacy.

When a RAT program has been run on a computer, it will install itself in such a way as to be active every time the computer is started subsequent to the installation. Information is sent back to the controlling computer (sometimes via an anonymous channel such as Internet relay chat [IRC]), noting that the system is active. The user of the command computer is then able to explore the target, escalate access to other resources, and install other software, such as DDoS zombies, if so desired.

A RAT will, therefore, consist of an installation function, a security bypass mechanism to provide access, a control function to provide domination of the affected computer, and an alerting method to inform the instigator of the RAT.

Distributed Denial of Service (DDoS) Structure

DDoS is a modified DoS attack. DoS attacks do not attempt to destroy or corrupt data, but attempt to use up a computing resource to the point where normal work cannot proceed.

The structure of a DDoS attack requires a master computer to control the attack, a designated target for the attack, and a number of computers in the middle that the master computer uses to generate the attack. These computers, in between the master machine and the target system, are variously called agents or clients, but are usually referred to as running "zombie" programs. The existence of a large number of agent computers in a DDoS attack serves to multiply the effect of the attack and also helps to hide the identity of the originator.

A DDoS client or zombie program has an installation program, an alerting function to inform the instigator, a listening function to await commands, and an attack function designed to deliver packets in the DoS attack itself. Note that this description applies only to the client program and not to the DDoS network as a whole.

Detection and Antidetection Techniques

While it appears logical to group detection and antidetection technologies together, they are not simply mirror images of one another. This brings us to an issue that every security writer must face at some point, and it seems appropriate to address it here.

Those new to virus research frequently assume that learning how to write viruses will help to protect against viruses. The old maxim of "set a thief to catch a thief" is so deeply accepted, that its application to information security is taken for granted. It is believed that learning how to break security automatically gives you an advantage in knowing how to protect against violations of confidentiality and integrity. While superficially reasonable, this argument is rather seriously flawed.

It is true that a touch of larceny in the soul is an advantage in the security game. Those who are completely honest and forthright expect everyone else to share the same code of ethics, and are, therefore, apt to be surprised by acts of deception. Successful security types tend to be the professionally paranoid: always on the lookout for ways to circumvent security measures, and therefore, means to prevent that circumvention. (Fortunately, most security mavens confine this tendency to the level of practical jokes. Combined with the adolescent emotional development of most technical people, this does explain the shenanigans at security conferences such as Defcon.) It is also true that a knowledge of the concepts of attacks can help in defense, in the same way that knowing the concepts of defense can help design better attacks.

However, teaching about the breaking of security systems is generally limited to specifics. Therefore, someone may learn about an express security vulnerability in a particular operating system or Web application. This will allow him or her to break into unprotected systems. An attacker may collect dozens of such vulnerabilities, as well as username and password combinations that allow him or her unauthorized access to resources or data. Knowing these individual attacks and accounts, a defender will be able to take steps to change passwords, close accounts, and patch systems to prevent those unique attacks.

Unfortunately, there are not dozens of possible vulnerabilities in systems, but thousands, and more are being discovered all the time. The defender who relies on this piecemeal approach to protection will be constantly struggling to keep up and will also be continuously at risk from those loopholes that have only just been discovered and not yet patched in the attacker's window of opportunity between those two events.

Eventually, though, one simply has to come to a basic conclusion: The skill sets involved in attack and defense overlap in places, but they are generally distinct. The most frequently quoted analogy is the one from Eugene Spafford: "Stealing cars and joyriding does not provide one with an education in Mechanical Engineering, nor does pouring sugar in the gas tank." Once you have accepted that people will try to deceive you, you do not have to develop particular skills in social engineering to be on guard against fraud. When you know that people are out to find user accounts with weak passwords, it is not necessary to know the inner working of password cracking algorithms in order to suggest that your users choose strong passwords. On the other hand, it is necessary for you to implement logging and auditing measures to be assured that all of your accounts are secure and remain so. Attackers have no such requirements for assurance.

To bring the argument more directly to malware, it is not necessary to learn how to write a virus to be able to defend against viruses in general. (In fact, the simple and very nontechnical expedient of not running unknown email attachments would protect you against approximately 95 percent of currently widespread viruses.) Over two-thirds of the viruses currently being written are created by modification of existing viruses, and this act requires only the most modest technical skill. In fact, even the programming of an "original" virus is a rather trivial task: Anyone with a knowledge of filesystem structures and file handling would be able to think of a dozen ways to perform the function.

At the same time, even the most skilled and creative virus writer would be hard-pressed to create a useful virus protection system. The attacker only needs to get lucky once to infect a system, and he can change targets if the task is too hard. The defender has to be right every single time, day after day, against a horde of attackers.

The broader implication to all of this discussion is that there is some level of danger in any writing or teaching about security. As noted, knowledge of defensive systems will assist the attacker in creating more effective intrusion measures. Some will say, in fact, that by writing about the techniques of software forensics, I will be alerting plagiarists and malware authors to their own signatures and allowing them to take measures to hide or alter those characteristics. That is always a risk, and every security author should be aware of the responsibility for balancing the benefits of what he writes against the potential harm of making the technologies known. In this particular case. I am only pointing out the myriad characteristics that can be used as identifiable signatures. Also, as will be noted in Chapter 8, changing your signatures is not always easy.

But perhaps we should turn to the more germane topic of detection and antidetection measures.

Detection Technologies

All antiviral technologies are based on the three classes outlined by Fred Cohen in his early research. The first type performs an ongoing assessment of the functions taking place in the computer, monitoring for activities known to be dangerous. The second checks regularly for changes in the computer system where changes should occur only infrequently. The third examines files for known code found in previous viruses. These antimalware technologies correspond to software forensic analyses: checking for specific operations emulation or code, determining through test runs the functions of a program, and looking for known characteristics. These basic types of detection systems can also be compared with the intrusion detection system (IDS) types, although the correspondence is not exact. An activity monitor is like a rule- or anomaly-based IDS. A change detection system is like a statistically based IDS. A scanner is like a signature-based IDS. After describing these main classes, we will look at some variations on the base technologies.

Activity Monitoring (Behavior-based)

An activity monitor performs a task very similar to an automated form of traditional auditing: It watches for suspicious activity. It may, for example, check for any calls to format a disk or attempts to alter or delete a program file while a program other than the operating system is in control. It may be more sophisticated and check for any program that performs "direct" activities with hardware, without using the standard system calls.

Activity monitors represent some of the oldest examples of antiviral software and are usually effective against more than just viruses. Generally speaking, such programs followed in the footsteps of the earlier antitrojan software, such as BOMBSQAD and WORMCHEK in the MS-DOS arena, which used the same "check what the program tries to do" approach. This tactic can be startlingly effective, particularly given the fact that so much malware is slavishly derivative and tends to use the same functions over and over again.

It is, however, very hard to tell the difference between a word processor updating a file and a virus infecting a file. Activity monitoring programs may be more trouble than they are worth because they can continually ask for confirmation of valid activities. The annals of computer virus research are littered with suggestions for virusproof computers and systems, which basically all boil down to the same thing: If the operations that a computer can perform are restricted, viral programs can be eliminated. Unfortunately, so is most of the usefulness of the computer.

In terms of software forensics, activity monitors will likely be most useful in the determination of the intent or operation of a piece of software under some kind of emulation. In more advanced analysis, the way that a program performs a given function may provide some evidence as to authorship.

Change Detection (Integrity Checking)

Change detection software examines system and/or program files and configuration, stores the information, and compares it against the actual configuration at a later time. Most of these programs perform a checksum or cyclic redundancy check (CRC) that will detect changes to a file even if the length is unchanged. Some programs will even use sophisticated encryption techniques to generate a signature that is, if not absolutely immune to malicious attack, prohibitively expensive, in processing terms, from the point of view of a piece of malware.

Change detection software should also identify the addition of completely new entities to a system. It has been noted that some programs have failed to do this and allowed the addition of virus infections or malware.

Change detection software is also often referred to as integrity-checking software, but this term may be somewhat misleading. The integrity of a system may have been compromised before the establishment of the baseline of comparison.

A sufficiently advanced change detection system, which takes all factors including system areas of the disk and the computer memory into account, has the best chance of detecting all current and future viral strains. However, change detection also has the highest probability of false alarms because it will not know whether a change is viral or valid. The addition of intelligent analysis of the changes detected may assist with this failing.

At present, change detection technology offers little of use to software forensics. In testing the intent or operation of unknown software, it may be helpful to do test runs on a baselined system and quickly determine changes that take place. Automated versions of change detection were implemented by the Victor Charlie antiviral package and by David Stang of the National Computer Security Association (NCSA; later ICSA, and now TruSecure). Extended forms of these systems may be used in the future to thoroughly test the functions contained in software.

The one exception to the lack of utility that change detection holds for computer forensics lies in the preservation of evidence. Cryptographically secure digests and hashes are vitally important for proving that the systems presented in court are equivalent to those recovered. At least one antiviral company has experimented with change detection systems that can be used to recover systems after infection has taken place. Broadly based change detection systems may be able to recover data even after erasure and may assist with logging and auditing operations.

String Search (Signature-based)

Scanners examine files, boot sectors, and/or memory for evidence of viral infection. They generally look for viral signatures—sections of program code that are known to be in specific viral programs, but not in most other programs. Because of this, scanning software will generally detect only known viruses and must be updated regularly. Some scanning software has resident versions that check each file as it is run.

Scanners have generally been the most popular form of antiviral software, probably because they make a specific identification. In fact, scanners offer somewhat weak protection because they require regular updating. Scanner identification of a virus may not always be dependable. A number of scanner products have been known to identify viruses based on common families, rather than definitive signatures. With enormous numbers of viruses now extant, signature scanners may become less popular, and we may see a resurgence of interest in the other generic detection technologies.

Signature scanning for viruses is somewhat related to determination of characteristic signatures that may be used for identification of software authors. Like virus signatures, author attributes must be determined in advance. Also like virus signatures, peculiarities of authors may need to be updated as they become known, and either imitated or modified.

Real-time Scanning

Real-time, or on-access scanning, is not really a separate type of antivirus technology. It uses standard signature scanning, but attempts to deal with each file or object as it is accessed or enters the machine. Because on-access scanning can affect the performance of the machine, vendors generally try to take shortcuts to reduce the delay when a file is read. Therefore, real-time scanning is significantly less effective at identifying virus infections than a normal signature scan of all files.

Real-time use of any of the base technologies may be of value in forensic analysis. Determining changes to a system as they occur, or the operations that the program under study attempts to perform, can provide helpful evidence.

Heuristic Scanning

A recent addition to scanners is intelligent analysis of unknown code, currently referred to as heuristic scanning. Note that heuristic scanning does not represent a new type of antiviral software. More closely akin to activity monitoring functions than traditional signature scanning, heuristic scanning looks for "suspicious" sections of code that are generally found in viral programs. While it is possible for normal programs to need to "go resident" to perform some operation, look for other program files, or even modify their own code, such activities are telltale signs that can help an informed user come to some decision about the advisability of running or installing a given new and unknown program. Heuristics, however, may generate a lot of false alarms, and may either scare novice users or give them a false sense of security after "wolf" has been cried too often. Combination systems involving both heuristic and signature (characteristic) scanning are likely to become basic software forensic tools.

Permanent Protection

The ultimate object for computer users is to find some kind of antiviral system that you can set and forget—that will take care of the problem without further work or attention on your part. Unfortunately, as previously noted, such protection has proven to be impossible. On a more practical level, every new advance in computer technology brings more opportunity for viruses and malicious software. As it has been said in political and social terms, so, too, the price of safe computing is constant vigilance.

For the foreseeable future, therefore, software forensics is likely to remain more of an art than a technology. Identifiable signatures and characteristics are not only still under development, but very much subject to debate.

Stealth and Antidetection Measures

Stealth is used inconsistently, even within the virus research community. By one definition, it is used as a reference to all forms of antidetection technology.

A more specific usage refers to an activity also known as tunnelling, which (in opposition to the usage in virtual private networks) describes that act of tracing interrupt links and system calls to intercept calls to read the disk, or perform other measures that could be used to determine that an infection exists. A virus using this form of stealth would intercept a call to display information about a file (such as its size) and return only information suitable to the uninfected object. This type of stealth was present in one of the earliest MS-DOS viruses, Brain. (If you gave commands on an infected system to display the contents of the boot sector, you would see the original boot sector and not the infected one.) Virus detection packages can check for signs of redirecting activity, or use change detection to determine whether system interrupts have been modified.

The tunnelling form of stealth technology will obviously get in the way of attempts to perform forensic analysis on software. It is unlikely that the analyst will be required to perform an assessment on a system where malicious software may be active, but such an attempt may be needed when, for example, an initial review is being done during the seizure of a computer.

Polymorphism (literally "many forms") refers to a number of techniques that attempt to change the code string on each generation of a virus. These range from using modules that can be rearranged to encrypting the virus code itself, leaving only a stub of code that can decrypt the body of the virus program when invoked. Polymorphism is sometimes also known as self-encryption or self-garbling, but these terms are imprecise and not recommended. Examples are the Whale virus and Tremor. Many polymorphic viruses use standard "mutation engines" such as MtE. These pieces of code actually aid detection because they have a known signature.

Polymorphism is probably confined to the world of viruses, but forms of encryption, both simple and complex, may be used by authors of software attempting to hide the operations of the program, text strings, or characteristic signatures.

A number of viruses also demonstrate some form of active detection avoidance, which may range from disabling of on-access scanners in memory to deletion of antivirus and other security software (Zonealarm is another favorite target) from the disk. As software forensic techniques become more widely used, it is likely that a variety of techniques will be employed to try to defeat the operation of forensic tools.

Summary

Viruses are not the only form of malware, and malware is not the only type of software subject to forensic analysis. However, virus research is the primary field in which software analysis has been performed outside of academic research.

While specific programming details are of limited use in protecting against malware, knowledge of the general concepts and historical means of attack can assist the defender in designing systems of protection.

The major components and basic structures of viruses and malware will also guide the analyst in determining what types of functions and operations to watch for.

Virus detection technologies point out the types of utilities that may need to be developed in order to assist in software forensic endeavors. In addition, the antidetection technologies that have been used in viruses indicate risks and pitfalls that may be encountered during program analysis.

7

Programming Cultures and Indicators

To those unfamiliar with programming, it may seem strange to talk about cultures in programming. However, you don't have to be around computers for too long before you realize that there are very definite communities, or trails of influence, involved in the development of programs and systems.

This is not as evident, perhaps, as it used to be. It is ironic to note that the availability of different kinds of programs is less nowadays than it was in the mid-1980s, for example. During the 1980s, I used approximately 40 different word processors in different situations. During the 1990s, I probably used four. Therefore, computer users formerly had much more of a chance to see a wider variation of programs in operation, and to see different types of approaches to essentially the same problem or issue.

User Interface

The most obvious difference in programming styles involves the user interface. In the current computing environment, not only does the Microsoft Windows operating system predominate on the desktop, but almost all programs present a very similar face to the user. (Speaking of history, note that this style of presentation does not, in fact, originate with Microsoft. The windowing style began with the Xerox Star computer system, and the standard system of menus was formerly known as IBM's Common User Access [CUA].)

Microsoft has, with its Windows operating system, made strong representations to developers to present a consistent user interface. This consistency has been promoted as a selling point, both to developers and users, through suggestions that similarity of style and presentation will reduce the need for the user to learn a lot about the programs before attempting to use them. In addition, libraries of code modules and application programming interface (API) tools have made this consistency easy to achieve—much easier, in fact, than developing new code for the same functions. This immutability gets carried too far at times, as with the near universality of a "File" menu on programs that have almost nothing to do with files.

It is, therefore, intriguing to note that different forms of display are now re-emerging in the concept of "skins," where popular programs can have add-on overlays that substantially change the appearance. The popular Winamp program, for playing audio files, presents a default appearance that mimics that of a physical stereo system, using buttons with the standard symbols for play, fast forward, rewind, and so forth. In addition, slider bars are used for volume control and toggle buttons for other functions, in much the same way the controls appear on a component system or boom box. This is in contrast to Microsoft's own Windows Media Player, with multiple menus and interface screens. Should a third such program come to prominence, it would be easy to see which program the author had used and learned from.

There is a cultural difference between UNIX and Windows in the user interface. When programs are invoked, they can have command line switches, or options. (This is true even of Windows programs: Look at the "Target" box on the shortcut dialogue and you will see that many options are called in this way.) UNIX programs traditionally employed a dash or hyphen to indicate these commands, whereas Microsoft generally used a forward slash character for the same purpose. The actual character chosen does not matter, and DOS or Windows programs can normally use both. When reading the documentation for a DOS or Windows utility, one can often determine the programmers' original background by the character used in command line examples.

More importantly, there is a cultural difference between the use of command line or keyboard controls and the Windows icon mouse/pointer (WIMP) graphical interface. Even in Windows programs, this factor may appear. Note that some Windows programs have very few keyboard shortcuts in relation to the menu options they present (Figure 7.1), whereas many others have a significant number of single-key commands (Figure 7.2).

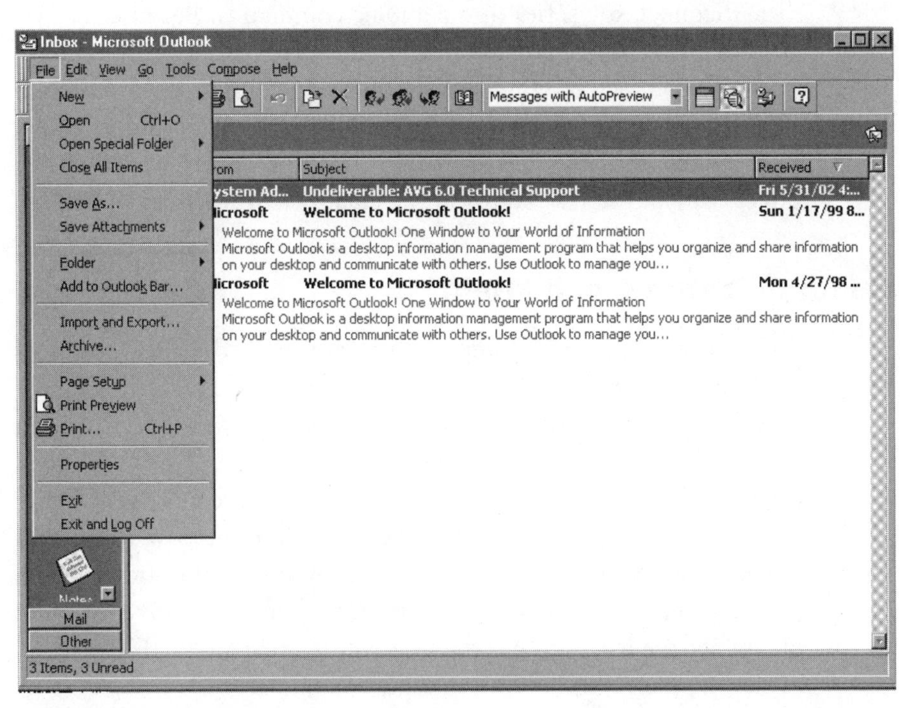

■ **Figure 7.1** *The Microsoft Outlook mail program for Windows. Note that out of 16 items on the file menu, only two have keyboard shortcuts.*

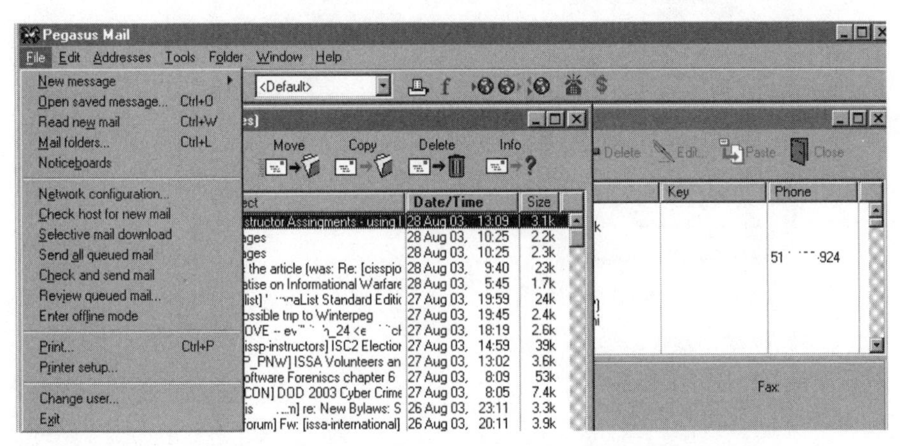

■ **Figure 7.2** *The Pegasus Mail program for Windows. Note that out of 16 items on the file menu, four, including the most important, have keyboard shortcuts.*

Some programs act as archetypes for an entire class of software. The Pascal programming language system developed by the University of California at San Diego was very popular in the early days of personal computing. The text editor for the system used a rather iconoclastic series of mnemonic commands. Because the system was used by many people who later programmed a number of the early commercial applications for microcomputers, those commands lived on for years in applications like WordStar and Sidekick. In the same way, the Perfect Writer word processor was almost a direct copy of the text editor in the emacs environment in UNIX.

Cultural Features and "Help"

Still, aspects of programming culture are more than skin deep (if you'll pardon the expression). For example, help functions in most systems now require that you know the exact term for the function that you are trying to perform. If, for example, I want to read (Figure 7.3), load, list (Figure 7.4), retrieve (Figure 7.5), get (Figure 7.6), or otherwise look at a file, I cannot find out how to do that under Windows Help. However, I have a number of options with a program that uses a help system that contains an index with synonyms for the features that the user wants, such as file (Figure 7.7), retrieve (Figure 7.8), and look (Figure 7.9).

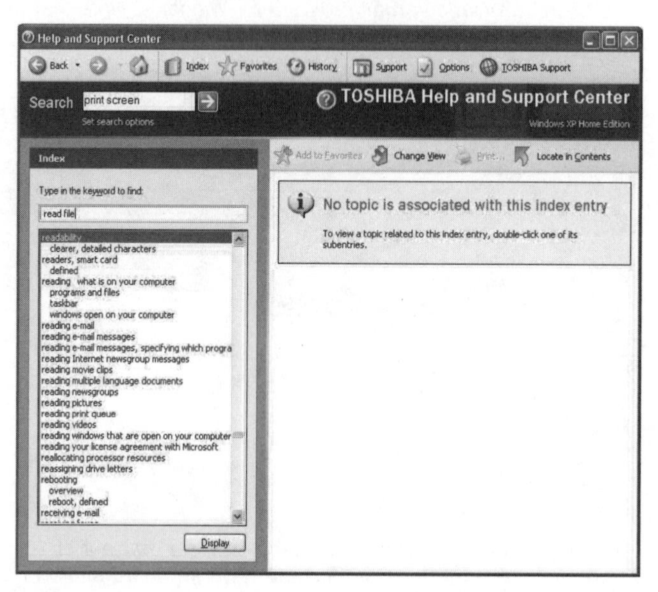

■ **Figure 7.3** *No help on how to read a file into the program or system.*

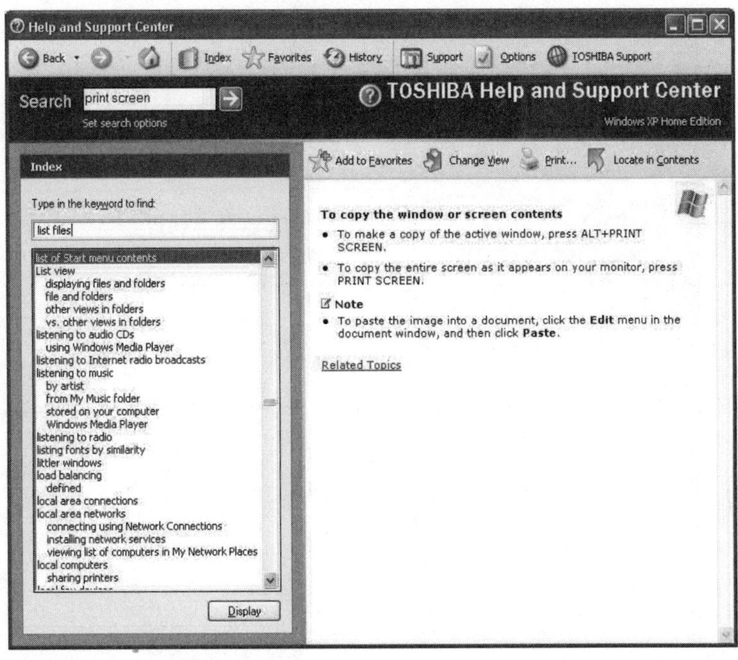

■ **Figure 7.4** *No help on how to load or list a file into the program or system (or look at a file, for that matter).*

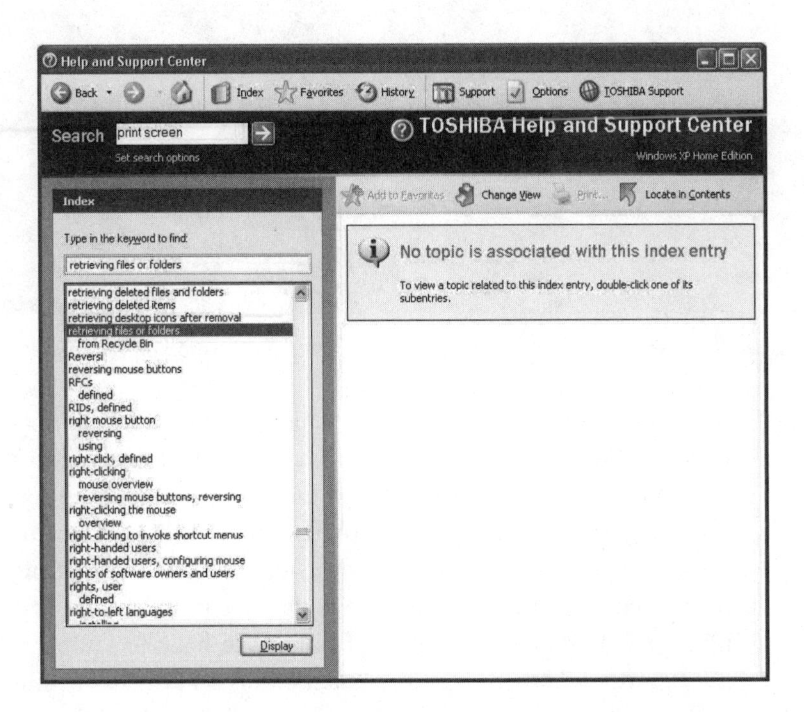

■ **Figure 7.5** *No help on how to retrieve a file into the program or system.*

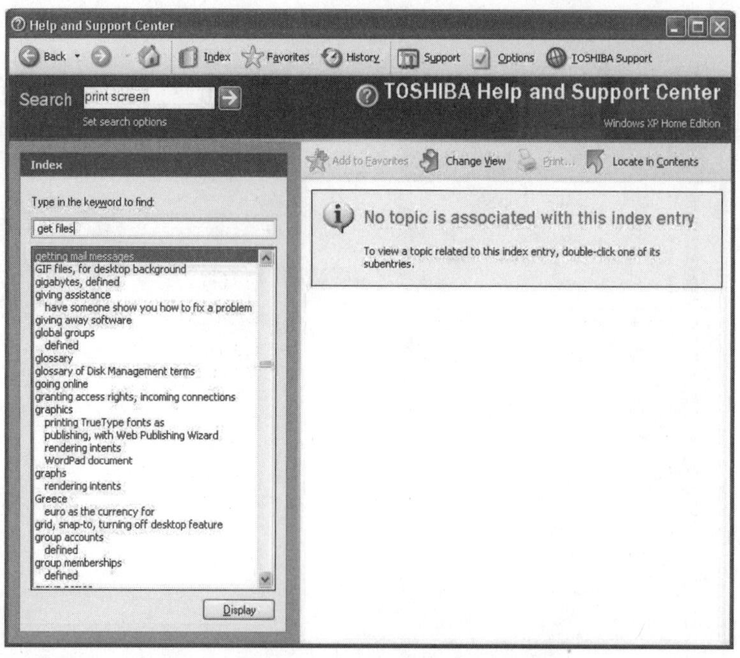

■ **Figure 7.6** *No help on how to get a file into the program or system.*

WP.EXE

Function Key	Feature	Key Name
F5	File Management	List Files
Alt -F6	Flush Right	Flush Right
Ctrl-F8	Font/Print Wheel	Print Format - 1
Alt -F8	Footers	Page Format - 6
Ctrl-F7	Footnotes	Footnote
F2	Forward Search	-> Search
Shft-F7	Full Text Print	Print - 1
Alt -F5	Generate	Mark Text - 6
Shft-F7	"GO" - Start Printer	Print - 4 Printer Control
Ctrl-Home	Go To	GoTo
Ctrl-F1	Go to DOS	Shell
Shft-F7	Hand-Fed Paper	Print - 4 then 3 Select Printers
Home-Space	Hard Space	Home-Space
Alt -F8	Headers	Page Format - 6

■ **Figure 7.7** *Pressing "F," for find or file, presents information on file management.*

Installation and setup programs provide another example. Everyone knows that you can speed through the installation of a program by simply hitting the "Enter" key on every screen and accepting all the defaults. An antivirus (and, oddly, change management) program called Integrity Master circumvents this mindless operation. Because the author wants to ensure that users make informed choices at a number of points, he has the setup program ask the user whether he or she understands what is happening. The default option is *not* to proceed with installation, but rather diverts

WP.EXE

Function Key	Feature	Key Name
Ctrl-F4	Rectangle Cut/Copy <Block On>	Move - 5
Alt -F5	Redline	Mark Text - 3
Alt -F5	Remove Redline/Strikeout	Mark Text - 6
F5	Rename a File	List Files - 3
Esc	Repetition Counter <n>	Esc
Alt -F2	Replace	Replace
Shft-F7	Report Printer Status	Print - 4 Printer Control
"-"	Required Hyphen	Hyphen
Home-Space	Required Space	Home-Space
Shft-F7	Restart Printer	Print - 4 Printer Control
Ctrl-F4	Retrieve Column	Move - 4
Shft-F10	Retrieve Text	Retrieve
Ctrl-F5	Retrieve DOS Text File	Text In/Out; List Files - 5
Ctrl-F4	Retrieve Text <Move key>	Move
Alt -F3	Reveal Codes	Reveal Codes
Shft-F2	Reverse Search	<- Search
Ctrl-F3	Rewrite Screen	Screen - 0
Ctrl-F3	Ruler Bar	Screen - 1

■ **Figure 7.8** *Pressing "R," for read or retrieve, presents information on retrieving or listing files.*

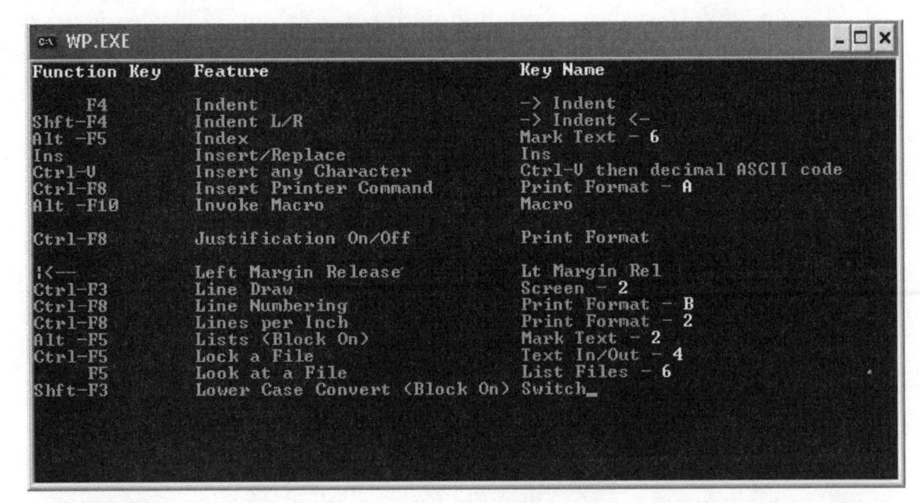

WP.EXE

Function Key	Feature	Key Name
F4	Indent	-> Indent
Shft-F4	Indent L/R	-> Indent <-
Alt -F5	Index	Mark Text - 6
Ins	Insert/Replace	Ins
Ctrl-V	Insert any Character	Ctrl-V then decimal ASCII code
Ctrl-F8	Insert Printer Command	Print Format - A
Alt -F10	Invoke Macro	Macro
Ctrl-F8	Justification On/Off	Print Format
|<--	Left Margin Release	Lt Margin Rel
Ctrl-F3	Line Draw	Screen - 2
Ctrl-F8	Line Numbering	Print Format - B
Ctrl-F8	Lines per Inch	Print Format - 2
Alt -F5	Lists <Block On>	Mark Text - 2
Ctrl-F5	Lock a File	Text In/Out - 4
F5	Look at a File	List Files - 6
Shft-F3	Lower Case Convert <Block On>	Switch_

■ **Figure 7.9** *Pressing "L," for load, look, or list, presents information on listing files.*

the user into a tutorial explaining the options (Figure 7.10). At the end of the tutorial, the same question is asked, and the same options are given. Therefore, if the user is "running on automatic," he or she will never get out of the tutorial. (So far, this does not appear to have spawned any cultural followers—but it should.)

The style of a help system is, of course, only one possible aspect of programming culture. The existence or absence of synonyms in a help file is far from being enough to identify a programmer (or pro-

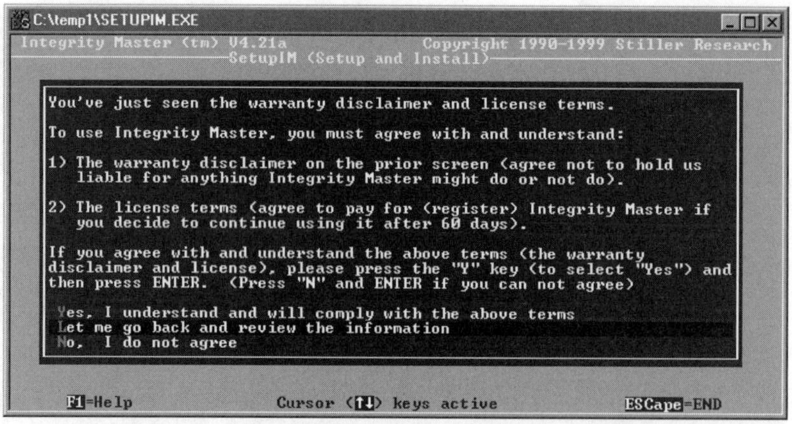

■ **Figure 7.10** *The installation program for Integrity Master. Note that the default option at this point in the program (dealing with agreement on the license terms) is not to proceed.*

gram code), or show a relation with some group. However, given a number of such indicators, it does become possible to show influences and relationships.

Functions

One source of indicators of cultural styles is the existence, or absence, of functions in the program itself. For example, a fairly universal function that used to be found in word processors was something called boilerplate. This was the ability to have a standard set of text, possibly a variety of frequently used paragraphs, and to import this text into the appropriate place in any document you were creating. The function was not always called boilerplate, but it was always there. Oddly, this function does not seem to exist in Microsoft's flagship word processor, Word. Of course, it is always possible to open another Word window, open the file that contains the text you want, select the text, cut or copy the text, close the second window or switch to the first, move to the position that you want the text to occupy, and then paste the text, but that does seem to be a rather involved process for such a simple function.

The absence of a boilerplate function, therefore, tells us that the original developers of Word were not thoroughly familiar with a variety of standard word processors, or, at least, were not familiar with actual word processing operations. In addition, if we then find another word processor that does not have a boilerplate function, we know that there is a very strong probability that the developers are primarily influenced by Word.

(This cultural indicator has influenced much more than a single program. The boilerplate function, and the ability to read a file into the entity you are creating, is present in many email programs. However, the ability is not present in Microsoft-based email applications such as Outlook.)

In dealing with program development and code, however, function has an additional meaning. Functions are not limited to the high-level operations that users see, but also involve lower level and "behind the scenes" transactions. There are, in fact, multiple levels of functions. These can involve decisions by the programmers about how much functionality to give users, such as the determination by Microsoft to provide programming tools in office productivity applications. There may also be principles governing the use, or avoidance, of lower level functions. For example, if you want a program to read a file, it is possible to make a call to high-level functions in the operating system to do so. It is also feasible to make low-level system calls. You might avoid the operating system altogether and make a direct call to the disk drive itself. It may even be viable to directly address the drive circuitry, providing very fine levels of control, but requiring intimate knowledge of the electronics. The decision as to what level of function to use will be based on the type of work that needs to be done, but will also be strongly influenced by the background and experience of the programmer.

Functions also refer to the structure of the program. When there is a need for a program to cycle through a procedure, a variety of options exist in high-level languages. The cycle can be programmed for a certain number of times, or may be dependent on a check as the cycle is entered, or may depend on the presence or absence of a condition, or may relate to the case of a certain value, or may even be determined by the process actually calling itself, as if it were a separate operation. Again, these options may be chosen on the basis of a specific need, but the selection will be strongly influenced on the basis of the programmers' level of comfort with a given structure, and this will depend on programs or groups with which he or she has worked.

Programming Style

Programming style can appear in many forms. There are large -scale issues of the approach to a problem, and there are minor technical details that may be telling in regard to the creator's background and inclination.

Program Structure

The overall organization of a program is one characteristic, and it is usually an indicator of a community. Programmers learn from those around them and absorb basic approaches without considering that there might be alternatives.

For example, a programmer from a UNIX environment (and that very often means an academic background, although less so than it did prior to the popularity of Linux) will try to use small utilities, and possibly lots of them. UNIX programs tend to have a single function and be adaptable to a variety of situations. Because of this extreme flexibility, effective use of UNIX requires a knowledge of the functions and outcomes of the utilities: They thus become used for purposes far from those originally intended. The UNIX cat utility, for example, was written to allow files to be tagged onto other files: The name "cat" is short for concatenate. Beginning UNIX users are often taught to use cat to read text files because the display device (typically the computer screen) is just another file to UNIX, and cat can be easily used to copy the file to the screen. Even the more "user-friendly" UNIX programs will manipulate these utilities for underlying functions. Many UNIX programs will, unbeknownst to the user, call upon an earlier, underlying application.

This preference and culture defines the nature of UNIX utilities. The programs are designed for piping, that is, providing the output of one program as input to another, so that a chain of applications can perform their specific functions on the data as they are processed through the pipe. As seen in the earlier boilerplate example, Microsoft applications are not designed even to conflate multiple existing files, let alone amalgamate the output of multiple programs.

(Some will point to Windows object linking and embedding [OLE], as a counter example. In fact, OLE is an example of the lack of cooperation between applications. When a section of data is cut from a Microsoft Excel spreadsheet and then pasted into a Microsoft Word document, only a pointer is created in the document. When the document is subsequently read, seemingly in Word, Excel is actually started in the background in order to process and display the spreadsheet data.)

In addition, DOS and Windows programs of the early 1990s were monolithic packages, having every function that the programmer could think of built in. The applications were very large, in terms of disk space, memory requirements, and the demands placed on the computer system. The programs were not designed to need, or allow, any assistance from other utilities. Almost all developers of commercial software for Windows followed this model.

Therefore, it was very easy to determine the background of a programmer from the overall design of the system. If it came in one large mass (and usually one large file), the person probably learned to program in a DOS or Windows environment. If the application was formed of individual functional modules, there was a good chance that developer had learned in a UNIX, probably college or university, milieu.

It is interesting and ironic to note that Windows programs are starting to emulate this pattern. Most Windows applications are now invoked by small files, which then make calls to dynamic link libraries. Thus, almost all programs share a common set of functions. This fact was used by Microsoft to argue that the Internet Explorer program was integral to Windows itself: Most of the code that the browser ran was also called by basic parts of the operating system. This organization is still not quite the same as the UNIX model because the UNIX utilities can be used as programs by themselves, whereas Windows libraries are simply collections of procedures.

The division or agglomeration of program files is only one aspect of the structural style of programs. Other factors may relate to the interface, such as the use of command line or menu options for the user. Some compositional aspects may be dictated by the project; for example, a multitude of separate programs that write out files for other packages to process in batches, as is needed in very large-scale enterprises. Other form elements may derive from the development tools. The software forensic practitioner will need to build an awareness of these styles and how they change over time. Such items may provide us with information both about the programmer's background and also about the intent of the code.

Programmer Skill and Objectives

When a programmer first starts out, it is a success simply to get a program to run at all. Considerations of logical design, clear flow of the code, and elimination of nonfunctional sections are less important. Therefore, the software of a neophyte will be marked by "spaghetti" code (where system calls are made as they were thought of, rather than in a logically designed structure), wasted space, bugs, and probably a number of self-congratulatory messages.

As the coder's skills develop, we will observe more structure and logic to the design of the program. The software will become easier to follow and understand (unless, of course, the programmer has deliberately obfuscated or confused the code of application). Bugs and irrelevant sections of material are likely to reduce in relation to the size and scope of the package.

At the intermediate level, the developer will begin to show preferences for certain other aspects of design and style. These endorsed factors may give us indications about either the background of the programmer or the intent of the software in terms of target or audience. The code may be preferentially designed to work on older machinery, limited memory, or to provide maximum performance in terms of speed. We must remember, at this level, to be careful about drawing conclusions not fully supported by the evidence. We may infer, from the fact that a program runs on older computers, that the author had only older machinery available. In the absence of other indicators, it may be equally likely that the programmer was intending to ensure that the software ran on as many computers as possible.

Older Computers

When speaking of older computer technology, we generally are considering the level of the CPU of the system. (We will address memory and space restrictions in the next section.) The series of Intel and compatible CPUs recognizable to users of Wintel machines have a core of opcodes in common. In general, any program that would run on an 8088 would run on an 8086, 80286, 80386, and so forth. Each new processor, though, generally had a few additional capabilities, and therefore, some extra opcodes that would not work on the older machinery. A program developed on a Pentium machine will not necessarily work on an older PC with an 8088 inside.

In general, programmers play with all the toys available to them. Therefore, it is reasonable to assume that a program that runs on older computers and has no requirements for the functions of the newer CPUs, was written by someone who had no access to modern machinery: a reasonable assumption, but not a safe one. It is entirely possible to restrict development to those operations that will run on all machines. This allows developers to appeal to the largest possible audience when selling software—or gives a virus writer the largest target population of victims.

We can draw an additional inference from the presence or absence of code requiring advanced processors. In cases involving intellectual property disputes, it may be that older base programming, containing all the real functions of the program, is plagiarized and then surrounded by a new interface. The absence of any opcodes that run only on newer CPUs in the fundamental operations, where the interface demands them, would support the contention that the essential programming was copied from earlier material.

Memory and Disk Space

This book is being written with WordPerfect version 4.2. Other than allowing those who are as old as I am a good laugh at the age of my software, this fact does have historical significance: The program was one of the last major commercial applications to be code-optimized. This means that the machine code itself was tuned and tweaked to conserve memory and to make the program run as fast as possible. In more recent commercial software development, these considerations are generally dismissed: Computers have enormous amounts of memory and disk space, as well as much faster processors, in relation to machines of a decade ago.

Programmers who are used to working within the confines of tight memory and disk space will have a tendency to reuse code in function and procedure calls, whereas those used to the larger space on modern machines will copy code as necessary, creating new modules that perform essentially the same tasks. In addition, more recent coders may not clean up sections of the code that have become redundant, frequently leaving large sections of software that are no longer called upon at all.

However, we can see evidence of such concerns at a much finer and more detailed level. An example, at the assembly programming level, of such tuning is to use the instruction XOR EAX, EAX to clear register EAX, and not use the command MOV EAX, 0, which would do exactly the same thing. The number of clock cycles required is two for both versions on the 80386 CPU. The former instruction requires two bytes of memory, but the latter requires five bytes, making it clear that this is a size optimization only: There is no benefit in performance, and the preference for using one over the other is strictly related to the amount of memory used in the program.

Conserving Cycles

Almost every good programmer aspires to have his or her application run quickly. Performance is a sign of experienced programming status, even in contemporary high-speed machines. Therefore, attention should be paid to attempts to squeeze every last cycle out of the computer. This will generally indicate a high level of skill and competence.

It may also indicate some familiarity with real-time programming, such as is used in embedded control machinery, or work with sophisticated and advanced projects, such as the competitions to crack encryption systems, or the distributed computation undertakings such as the analysis of space radio traffic in the search for extraterrestrial intelligence.

Examples of highly tuned programs will probably be indicated only at the machine code level. Truly optimized code demonstrates assembly language programming at the very least, but we must be careful: Most optimizing compilers will make some substitutions in the name of performance.

Elegance

As programmers attain truly elite standing in their craft, mere technical signatures become difficult to find. Eventually, coders bring a kind of artistry to their work. The program structures, algorithms, and shortcuts achieve what mathematicians refer to as elegance. This type of characteristic is hard to describe, and possibly even harder to pin down. Ultimately, it will be recognizable only to a programmer near the same level of skill.

Developmental Strictures

As noted elsewhere, we need to be careful, when developing evidence in regard to programmer identity, that we are not simply identifying the tools that are used. Developer utilities, applications, and environments all influence a number of aspects of the final program, and, while these characteristics can provide us with some information about the programmer, such as a preference in tools, we need to determine which evidence is produced by the craftsman, and which by the instrument.

As noted, compilers can create their own signatures. Some programs also will be asked to optimize either for memory space or for speed. Related tools, such as libraries or application programming interfaces, may also introduce signature code or styles that have nothing to do with the programmer him or herself.

Other tools, such as text editors or programming environments, may subtly influence the complexity of functions or statements, and even aspects such as variable names. The preference of the original programmer for such a utility may, therefore, influence a whole range of subsequent coders.

More advanced vehicles, such as computer aided software engineering (CASE) tools, may bias the entire structure of a project. Many of these applications provide project management capabilities, and the assumptions of the precursor developers will drive all aspects of the development enterprise.

Technological Change

The software forensic specialist will need to develop information about indicators and signatures from reviewing and using many different programs. Over time, another factor will become obvious: Characteristics change as development technology changes. In addition, the dominant programs in the market will rotate at intervals, and this can be both a help and a nuisance in developing evidence. In 1990, indications of a UNIX culture would signify a probable academic background, whereas nowadays, it would more likely mark a hobbyist. In 1990, experience with the common user access (CUA) displayed IBM experience, now it is common in the Windows world.

Summary

As with any other human activity, there are societal influences that incline programmers toward or away from certain patterns. These factors may result from groups, communities, training, or simply the availability of patterns to emulate. Cultural inspiration can be identified in software and can provide some indications that may assist us in trying to identify a particular author. At the same time, we need to ensure that we do not mistake common traits for specific signatures of individuality.

8

Stylistic Analysis and Linguistic Forensics

My approach to the electronic fingerprinting of fraudulent letters arises from work related to the collection of forensic evidence from computer-related activities. Other fields of study have contributed approaches that must not be neglected. Particularly with respect to source code analysis, work that has been done on literary analysis, and particularly on the authorship of disputed documents, is a much more mature field. In the case of Biblical criticism, this work literally goes back centuries.

The contribution of stylistic analysis and linguistic forensics obviously will apply more to authorship identification and possibly detection of plagiarism and intellectual property issues than to the concerns of intent and analysis of malware. In addition, it may not always be immediately obvious just exactly what the application of a particular technique of textual analysis contributes to software forensics.

Stylistic forensics may be performed on source code. Code is, of course, more structured than free-form writing, and definitions of sentences may be problematic in relation to command statements, but the approaches would be fairly straightforward. Comments could be analyzed with little or no modification of techniques that would normally be applied to a manuscript. We have previously noted that text strings may not always appear in object code. There are analogies that can be made between writing and software. If one considers opcodes to be letters, then by extension, functions and operations may be words, and modules might be considered sentences. If we understand the underlying concepts of stylistic analysis, they can be utilized in software forensics.

While identification of the canon of the Christian Bible and the authors of individual works is probably the oldest arena of such work, a number of classic cases of research should be familiar to anyone. The most widely known is probably the debate about the authorship of those plays attributed to William Shakespeare. Many doubt that these seminal works of English literature were written by an otherwise unknown actor from the town of Stratford. A number of candidates have been proposed, chief among them the poet Sir Francis Bacon. (Interestingly, in view of the fact that we are now dealing with technical pursuits, another contender is the early scientist Roger Bacon.) The authorship of a number of literary works is open to question, and there have even been attempts to discern the contributions of sundry individuals to the formulation of the Constitution and Declaration of Independence of the United States. The techniques applied to all of these inquiries can be equally applied to source code analysis. In some cases, the strategies can be applied to the investigation of object code.

Biblical Criticism

The investigation of which writings were appropriate to include in the Christian Bible dates back to at least the fourth century, when the current canon was roughly established. Even today, the canon is not absolutely defined. "Study" Bibles will include footnotes to the effect that some sources include additional verses or slightly variant wording. There is a major division regarding the canon between the Catholic and Protestant churches: Catholic Bibles, such as the Jerusalem version, include works of the Apocrypha, the "hidden books," such as Tobit, Judith, and the Maccabees (including an additional chapter of Daniel that contains early instructions on forensic interviewing technique).

Biblical criticism technologies can be applied to other materials as well. Two of the recent refinements and clarifications of older methods are form criticism and redaction (or editing) criticism. Each of these procedures can contribute to the understanding of a work.

Form criticism considers standard literary forms found in the literature and stories of the time. In the historical age when the Old

Testament books of the Bible were being collected, common forms were creation stories, miracle stories, and wisdom stories. These forms, and others, appear in the Old Testament. By comparing the examples of forms that are part of the Bible with those in other literary works of the time, similarities and differences can be noted. The existence of the form-based stories themselves, and variations from the norms found in other works, provide significant evidence about the formation of the books that comprise the Old Testament, including indications of authorship.

Redaction criticism also notes the standard forms, stories, and other component pieces of a work. In redaction criticism, though, it is the editing and structuring of the constituent parts into a whole that is of greatest interest. The modification of an original essay known elsewhere, and the architecting of the overall creation, provide information about the intent and possibly identity of the editor.

Recall from our discussion of decompilers in Chapter 4, these two types of higher criticism correspond almost exactly with the two classes of decompilation. On the one hand, we are looking for atomic and self-contained forms, whether they are literary story types or programmatic functions. On the other hand, we examine the structure into which the components are placed. Thus, the oldest form of literary analysis relates to the basic approach of the newest form of forensic science.

Shakespeare and Other Literature

The debate over the authorship of Shakespeare's works and plays is obviously fascinating to a large section of the population: Everyone seems to be in on the game. (No less a literary luminary than Mark Twain pointed out that Shakespeare's parents were illiterate, his will was that of a businessman rather than a poet, his will doesn't mention any literary works, and only one manuscript poem can be compared with the handwriting in his will.) In fact, if you try to do Web searches on "authorship analysis" or similar phrases, you will find that sites concerning Shakespeare comprise the bulk of the references you get in return.

In reading the various discussions, you will note that the same types of criticism apply. A specific story form dealing with an alchemical or philosophical point, such as may be found in a play like "The Tempest," will be used to argue for authorship by Roger

Bacon. A particular poetic structure will be evidence to promote Francis Bacon as the creator of the Shakespearean canon. The exact items may vary, as well as the candidate authors, but the techniques are basically the same. (To avoid adding to the debate, and to return my readers to the topic of this book, a significant computer analysis of Shakespeare's works indicates that they are consistent enough to make the idea of a committee infeasible, and distinct enough from the known writings of proposed contenders to make claims for their authorship unlikely.)

The Federalist Papers are a set of documents written by some of the framers of the American Constitution (Jefferson, Madison, and Hamilton are frequently credited), but for whom precise attribution is not given. One famous analysis used statistics, only to find that the traditional literary investigations based on context and vocabulary did not work. One of the only aspects of the text that did give evidence of authorship was the use of connective words. Other statistical analysis on the works of Shakespeare, based on what are called type-token relationships, provides at least superficial indications that the works are not by a single author, even when different types of literature and plays are considered separately.

Frequently, literary analyses directed at authorship compare the use of certain vocabulary items previously identified as typical of the candidate authors on the basis of known works, the use of certain grammatical constructions, and the counts of certain grammatical types such as the ratios of nouns or adjectives to verbs. Others may look at word-length spectra (and a number of statistics calculated from them) or sentence-length spectra. Note that the first group of factors relates to the content of the text itself. The word- or sentence-length calculations do not seem to have any relation to the content or topic of the text itself. This division will become important in our discussion of the different types of approaches to text analysis.

Many of the arguments in Shakespearean and other research will be advanced by dedicated scholars, who have studied the subject for many years and have a great deal of knowledge of the subject overall. Note, though, that the popularity of the Shakespearean game provides for a mass of proposals from enthusiastic amateurs. Reading through this mountain of rather ill-informed discourse does have one benefit—it points out a significant error to which forensic linguistic analysis can fall prey—over reliance on content analysis.

A lot of the reasoning proposed in literary debates is based on content-related analysis. This would pertain to the existence or use of certain vocabulary, or characteristic turns of phrase. It might correlate with the use of specific adjectives, certain verb forms, analogies, or verbal images. Content analysis, as we noted in the chapter on legal issues, is compelling, and presents itself as "obvious" to an analyst—particularly when other forms of evidence are hard to find.

Content analysis is generally based on the first of the three forms of identification authentication—something you know. Somehow, we find this to be very convincing as evidence. When discussing authentication, we automatically think of the classic "something you know" example, that of passwords. So the "something you know" aspect of content-based analysis appears, to humans, to be of a higher standard of proof than other forms.

Content-based analysis can set us up, as analysts, to deceive ourselves. I recall a book of some years ago that set out to prove Karl Marx was a Satanist. It used passages from Marx's own writings. Presented in isolation from their original content, and aggregated in support of this thesis, the excerpts did present a very compelling argument that Marx held to a satanist philosophy and theology. Of course, when examined in context, and in light of Marx's writings as a whole and in relation to the literature of his day, this proposition became absurd. The melodramatic passages about darkness reflected the idiom of the times and thoughts about the state of society, rather than personal philosophy.

Content-based analysis can, therefore, be deceptive. While it can provide us with some evidence and insight, as we noted in Chapter 7, we should be reticent to accept it as proof in the absence of other indications. We can turn to the other authentication factors. "Something you have," in the case of software, will probably not be very helpful. "Something you are," though, turns out to have a very good record. Both writers and programmers generate characteristic signatures in creating text and software. Noncontent signatures are very dependable. We will examine both content and noncontent analysis, and one particular form of noncontent analysis, later in this chapter. Evidence indicates that noncontent analysis can track identify over long periods of time, even in the face of deliberate efforts to change the means of expression.

Individual Identification and Authentication

We may be fortunate enough, in examining software, to have specific and explicit references to an individual. At first consideration, this would seem to be a good thing in terms of identification. Remember, though, that identification is only asserted by the user. In any security situation, identification must be authenticated.

The three standard forms of authentication, as noted briefly in the preceding section, are something you know, something you have, or something you are. The classic example of something you know is a secret password, which must be properly paired with a specific username. Something you have may be a physical key or an access token. Something you are is generally some kind of biometric measurement, such as a fingerprint. However, another kind of biometric is a specific behavioral characteristic, such as a voiceprint, pattern of motion in creating a signature, or speed and sequence of typing.

Content analysis, as we have seen and shall see, can be used to verify the identity of an individual. If a piece of information is known only to one individual, then that can authenticate the person. Other than in a properly established password system, the limiting of the information to one personality makes it difficult to use for authentication. However, content analysis can gather evidence of a cultural nature. This is roughly analogous to the situation in World War II when German soldiers would be dressed in American uniforms and sent to infiltrate American units, only to be detected by Americans querying them about baseball scores and pennant winners.

Knowledge can be either specific or cultural, and it can also be either explicit or implied. Certain types and styles of programming, or certain approaches to a problem, can provide an indication about the knowledge of the programmer. In a piece of text supposedly written by a European of the Middle Ages, knowledge of potatoes and maize corn would be out of place. (Those crops are not native to Europe and were imported from North America sometime after the voyages of Columbus.) In the same way, aspects of object-oriented programming would be out of place in software that was supposed to have been written in the early 1980s, and would support the contention that the developer created or revised it at a later date.

Something you have is rather problematic in the programming world. If someone were to use a completely idiosyncratic compiler, it might create a unique signature that could be used as a means of authentication. (On the other hand, that signature could also become known and copied.) Microsoft attempted to create something similar with the global user ID (GUID). This was supposed to be unique to a specific user, particularly when it was formed from a unique hardware identifier, such as the media access control (MAC) address of a network card. Unfortunately, the GUID, once known, can be forged, copied, altered, or eliminated from objects that should contain it.

The compiler is not the only object that can affect the development of a piece of software. Many tools may contribute to the creation of a piece of code, and all may leave some kind of signature. Development "environments," editors, libraries, computer aided software engineering (CASE) tools, and other utilities can all contribute information about something that the programmer had.

The term "electronic fingerprint" is particularly well chosen in regard to identification of individuals. Physical fingerprint evidence frequently does not help us identify a perpetrator in terms of finding the person once we have a fingerprint. However, a fingerprint can confirm an identity, or place a person at the scene of a crime, once we have a suspect. In the same way, the evidence we gather from analyzing the body of a program may help to confirm that a given individual or suspect is the person who created the software.

A physical fingerprint is one example of biometric identification, a form of identity authentication that relies on something that the individual is. In the programming process, the person does not leave physical trace evidence because software is, after all, only ones and zeroes. Biometrics are not confined to purely physical quantities, though. A number of devices and products on the market measure signature dynamics, voiceprints, typing patterns, and other behavioral characteristics that can be saved in digital formats.

To be sure, once a file has been created and saved, there is little evidence to suggest how fast it was typed. There are other indications that do survive. In terms of text and software, it is difficult to make a hard and fast distinction between what a person knows and what he or she is, but both can be used to identify or verify the identity of the author.

One fairly obvious means of identification lies in errors. As a typist, I am continually typing "teh" instead of "the," "adn" instead of "and," and, for some completely inexplicable reason, "typoing" instead of "typing." The relative rates of these errors would probably form some kind of biometric measurement. In terms of software forensics, it is unfortunate that such simple typing errors are often removed in the course of the programming process. An error in typing a command becomes a fatal error at compilation time, and is therefore corrected, in a kind of parallel to the automatic spelling corrections that go on in modern word processors. Still, some kinds of characteristic errors, such as a preference for an inappropriate class of control structure, may survive into the final program.

Handwriting analysis also involves identifying the properties of a writing sample. A feature of the script can be anything identifiable about the physical penmanship. These marks are usually common variables, such as the way the i's are dotted or average height of the characters. The traits useful in handwriting analysis are the writer-specific features, but they do not need to be unique. An attribute is said to be writer specific if it shows only small variations in the penmanship of an individual and large variations in the writings of different essayists. Features considered in handwriting analysis include the shape of dots, relative proportions of lengths of parts of the letters, the shape of loops (such as narrow or broad), horizontal and vertical extent of the writing, slant, and regularity, and fluidity of the flow of strokes. Most features in handwriting show little variation. However, most writing will also contain features that set it apart from the script of other authors: attributes that to some degree are unusual. A sample that contains i's dotted with a single dab probably will not yield much information from that characteristic. However, if all of the o's in the sample have their centers filled in, that feature may identify the author. Still, an accumulation of minor variations may also provide identification. In the same way, a collection of programming signatures that show only minor deviation from the norm may, in aggregation, contribute to the identification of a programmer.

(Note that handwriting analysis, concerning the identity of an author of a particular piece of handwritten material, significantly differs from graphology. Studies indicate that the attempt to discern personality traits from handwriting does not appear to have any predictive value.)

Both the content and the syntactical structure of text, source code, and object code can, therefore, provide evidence that relates to an individual.

Content Analysis

Recently, my wife drew my attention to very similar embroidery charts in a book from her library and a handout given away by a company that manufactures embroidery floss. The composition, proportions, and detailed parts of the images were identical. The obvious conclusion is that either one copied the other, or that both were copied from a third source. If this situation were to be examined in terms of intellectual property, and in the absence of any evidence of a third source, we could note that the window framing in the free pattern is more complex, and that the free pattern has additional shading around the window. This would tend to indicate that the free pattern had been copied or modified from the book because copiers generally are predisposed to embellish, rather than simplify.

This is another example of the content analysis we discussed earlier in relation to literary analysis. The charts provide a specific representation of an idea, in this case presented in graphical form. In the same way, an idea presented in text will have a certain representation, composition, and structure. The way that authors present ideas in text tends to be characteristic, both in terms of overall composition, and in terms of details such as vocabulary and phrasing.

In addition, we can use analysis of text to find sequences of messages and trace influences. In texts that are copied from an original, the overall structure and composition is usually unchanged, but details and embellishments are often added. In terms of works such as the Bible, we note that copyists incline toward "correcting" obscure passages in later versions, often to conform with the prevailing philosophy of the day.

The syntax of text tends to be characteristic. Does the author always use simple sentences? Always use compound sentences? Have a specific preference when a mix of forms is used? Syntactical patterns have been used in programs that detect plagiarism in written papers. The same kind of analysis can be applied to source code for programs, finding identity between the overall structure of code even when functional units are not considered. In Chapter 4, we examined a number of plagiarism detection programs that are available, and the methods that they use can assist with this type of forensic study.

Of course, when considering the content of the text, most people consider characteristic use of vocabulary and phrases. This does tend to be effective, but it usually relies on having a large set of samples to analyze. Generally, we must ensure that the texts cover the same, or similar, subjects to avoid problems with disparate vocabularies in differing fields.

In terms of software, source code is quite restricted in regard to vocabulary. There are the keywords or commands of the language, and they allow for no variation. There is choice available in the use of variable names and commentary, but often these choices are heavily influenced by schools, groups, and other aspects of programming culture, as we noted in Chapter 7.

The vocabulary of software is generally the construction of the functions themselves. In this regard, individuals follow known and comfortable patterns, at least known and comfortable to themselves. If we examine this kind of vocabulary, we will find expressions that may identify an individual.

Along with the problem of the deceptively compelling nature of content-based analysis, there is an additional factor. The content and style that are used by the author may also be readily recognizable to the author him or herself. Therefore, it is much more likely that an artist intent on hiding his or her identity will be able to alter patterns in the content of the material. This problem exists in software development, as it does in literature, but modification of style may be more difficult for the programmer than for the author.

Error Analysis

Errors in the text can be extremely helpful in our analysis and should be identified for further study. In some of my early work published on the history of computer viruses, I made a mistake in the spelling of the name of one person involved in the creation of a specific program. Shortly thereafter, another person also published such a history. The histories were very similar, but that could be expected if two people both had access to the same sources. However, the second history also contained the error that I had made. The author of the second history, had he followed original reference materials, would not have made that error, thus indicating that the later text was a copy of my original. Hence, the detection of plagiarism is made much easier. Producers of reference works, in particular, often deliberately introduce minor errors to their work to determine whether the products have been copied.

In another example, two students cheated and copied answers on a test written for a course I gave at a college. In my report on the incident, I included a statistical analysis on the results. The likelihood of two students getting exactly the same questions correct was extremely small. But the chance of two students making exactly the same errors was five times smaller, making a much stronger case for the cheating hypothesis. Again, minor errors may not be convincing individually. When aggregated, however, we may be able to present a very strong case based on the unlikelihood of a proliferated pattern of errors.

In terms of software forensics, error analysis can be very helpful indeed. Very few programs are completely free of bugs. (It has been said by the eminent Edsger Dijkstra that if debugging is the process of removing bugs, then programming must be the process of putting them in.) While spelling and syntactic mistakes will be detected by the compiler and removed by the programmer before the system can be completed, errors in logic are not. When marking student programming assignments, I often find that the earliest indication of duplicated or plagiarized programs is a characteristic pattern of errors in the results of two programs. Students will change internal comments, variable names, the order of modules, and sometimes even the user interface, but will fail to debug the logic of the code they copy.

In virus research, we saw—and continue to see—the same phenomenon. We are able to trace families of viruses due to similar features and coding, but the real proof that someone has simply copied virus code, with minor modifications, lies in the mistakes that are made. Fortunately, viruses and other malware are absolutely riddled with bugs and provide a rich field for tracking the origin and sequencing of versions of malicious programs. If a bug is corrected, this is often an indication that the original author is behind the subsequent edition. The number of errors is usually sufficient that we can track the changes made through half a dozen or more variants of a virus by the initial author. (The correction of mistakes is one of the indications used to place variants in sequence within a family.)

Noncontent Analysis

My wife once worked as a secretary in a government office, typing reports for a variety of officers. She noted that different writers had different styles. In those days of typewriters and monospaced fonts, one of the factors she discovered was that dictation from different people required different line lengths. If the wrong length was used, the report turned out to have a ragged edge down the right side of the page. If the proper line length was used, the margin was neater. The line length was characteristic of the writer: Tom needed a 65 space line, Dick used 72, and Harriet required 68 spaces. This characteristic is consistent over time: Harriet will always need 68 spaces. This may seem to be a trivial characteristic, but it does indicate that a number of identifying attributes are available to build an electronic fingerprint of text.

These characteristics do not appear to relate to the content of the text being written. Like biometrics, they seem to be a factor of who you are, and not what you are discussing. Thus, they are more dependable than content-based measures such as vocabulary, which will be modified by the topic of discussion. In addition, authors are generally not aware of noncontent factors. This means that an author, even when attempting to disguise his or her style of writing, will find it much more difficult to modify traits of which he or she is not cognizant.

As previously mentioned, noncontent signatures do not seem to correspond to the subject of the text. While there is good evidence to indicate that this is true in many cases, there is work yet to be done in this field. In another example from personal experience, my wife has found that noncontent characteristics of my own writing, such as the line length factor noted above, do vary depending on whether I am writing technical material, expository pieces, or humor. Therefore, until such time as more is known about these factors, noncontent measures should be chosen and studied carefully in relation to their use in forensic situations.

Cusum

A specific method of finding such noncontent characteristics is "cusum," invented by Andrew Morton, and originally applied to higher criticism of Biblical texts. It is thoroughly explained in the book *Analysing for Authorship*, by Jill M. Farringdon.

Literary critics are quite used to talking about how an author such as Henry James would write enormously long sentences, sentences that would, in more modern writings, be split into smaller, more digestible chunks, but which were, in the days when it was considered acceptable for someone like Marcel Proust to write an entire book that was one long sentence, the norm that was to be emulated and adopted. Others wrote differently. Hemingway, for example. Short sentences. Sentence fragments, really. So critics are quite used to making decisions about authorship based on numeric metrics.

Cusum (or QSUM; the two terms seem to be used interchangeably in the Farringdon book) is such a technique. Instead of looking at meanings or characteristic turns of phrase, the method looks at combinations of statistical patterns in writing, patterns that the writer is probably unaware of using.

The bases of comparison are generally sentence length in proportion to the number of short words, and words starting with vowels. A number of these factors can be used in a variety of combinations. Depending on the author, the amount of text available to study, and possibly the type of literature being analyzed, different combinations are more or less effective in discerning authorship. This use of seemingly meaningless metrics may sound strange in regard to the identification of an author. An analysis of general word use in English indicates that cusum is based on syntactic structures, rather than content. Thus, what cusum may be measuring is the characteristic preferences that an author may have in terms of syntax, rather than vocabulary. Again, this relates to the division that we saw earlier in the chapter on advanced tools and decompilers, determining the content in terms of logical operations and functions of a program, or divining the overall structure and how the components fit together.

The statistical fingerprint of authors remains consistent over time, along with the development of writing skills and style. This constancy is important when dealing with matters of identification and authentication. It is possible to have biometric identifiers that uniquely distinguish an individual—but which change over time. Prior to the establishment of physical fingerprinting for verification of identity, the system most widely used was one called *bertillionage*. This practice took measurements of the fixed or long bones of the body and employed the aggregated values to determine identification. (Height was not one of the metrics that could be used because the total length of the body can vary due to spinal curvature and compression of the disks: Some people are able to alter their height over a range of seven centimeters.) Bertillionage could not, though, be used with children, because their bones were still growing. In the same way, textual analysis based on content measurements, such as vocabulary preferences, will change over time as the writer develops knowledge, writing skills, and experience. (Vocabulary, when compared against the writer's age group, turns out to be a very quick, and for simple situations, remarkably accurate assessment of intelligence.)

In regard to noncontent text signatures, and cusum in particular, research has demonstrated that patterns and signatures remain identifiable and reliable, even when the author notes a significant change in his or her literary "voice." Such a change in voice is partially conscious, but is also partly unconscious, and it is very telling to note that even deeply rooted transformations do not void characteristic signatures.

In another study, an editor's style was identified, despite an attempt made to represent a piece as solely the work of a single author. In cases of editing, it may not be possible to come to a final and incontestable determination of authorship, but some evidence can be obtained. In any case, the very fact of evidence of editing may be important. The problems of editing may or may not be a factor in collaborative works. In some situations, it has been possible to use cusum to determine authorship even on a sentence-by-sentence basis.

In a forensic case, cusum was used to prove that a transcript of an interrogation was not a transcript, but had been edited. Cusum can be used with verbal, as well as written material, and can be used to determine whether someone for whom written samples are available, verbally recited a piece that was attributed to him or her. This is truly impressive for anyone who is familiar with the extreme difference between written and spoken material.

Consistent patterns survive in authors from childhood into adult life. Where writing samples for an individual are available over time, cusum has been able to find signature characteristics that appear in childhood, the teen years, early adulthood, and senior years. Ironically, other studies have shown that the use of complex sentence structures in youth may be a reasonable indicator of lowered risk of Alzheimer's disease much later in life. Whether these factors are correlated has not yet been determined.

Not even the use of dialect and invented languages can hide an author's signature. Cusum has been studied in relation to efforts by experienced writers of fiction who are used to distinguishing between characters by clear and obvious differences in style, sentence structure, vocabulary, and faulty grammar and spelling. In addition, trained philologists and linguists have created entire languages, with artificial but realistic vocabulary, grammar, and syntax. In all cases, cusum has been able to find signatures despite these deliberate and extreme attempts to disguise the identity of the author. This success makes it extremely unlikely that deliberate deception would be able to withstand noncontent analysis.

Speakers and writers of English as a second language have remarkably consistent signatures over time. Again, this is intriguing, given the fact that such people have already learned one syntax and are having to modify their thinking based on a new one, but it is not surprising in light of the fact that children have constant characteristics despite their own continuous language development.

Cusum is neither perfect nor a panacea. Its practice is still more of an art than a science. In particular, there are a number of areas of study that have not yet been completed, such as overall statistics on the ability to make distinctions on the basis of these metrics over a large population. These studies would probably strengthen the process, and make it more useful and accessible.

The technique has had considerable success in a wide variety of situations, but still does not have the backing of statistical surveys indicating the strength of identification when it has been made. It is a truism in regard to physical fingerprints that they are unique to the individual: like snowflakes, no two are alike. Less well-known is the fact that the standard means for classifying, searching, and comparing fingerprints does not provide unique identification. The odds of any two random individuals matching is approximately one in six million. This means that a given set of prints will match five people in the country of Canada, or more than eight in the state of California. Similar figures can be obtained for DNA matching and other forms of biometric identification. It is important to know how precisely, and how accurately, identification can be made. It is equally vital to be able to state what the chance of an error is.

The same statements cannot be made for cusum. How many authors must we analyze before we find two with indistinguishable signatures? If we find a match with one characteristic, how likely is it that a second pattern will match as well? Which metrics are most reliable for determining authorship, and how many metrics, considered together, should be the basis for a reasonable identification? Even within a specific characteristic, certain patterns can be more useful than others. Returning to the field of physical evidence, finding an AB-negative blood type limits the possibilities to only 15 out of a thousand people in a population. Finding an A-positive sample is not as useful for identification because about one-third of people have that blood type. Work needs to be done to fill out our understanding of these issues.

One of the strengths of cusum when presenting in court is that the preferred method of analysis is graphical. A graph can be plotted showing writing samples by two different authors, and it will present a clear discontinuity. Another plot will trace a known sample by our putative author and the text under consideration, and, lo and behold, the curve is smooth and has no jumps. Cusum can also be used with statistical appraisals, but these require an understanding of the mathematics involved and are not considered as compelling when presented in court.

This strength, though, can also be a weakness. The choice of graphs to present in evidence is a bit of an art. Some metrics are more dramatic than others, and the choice of scale and origin for the plots can have a significant impact. It is quite possible that a knowledgeable opponent would be able to take your base results, present them in court with an exaggerated scale, and, at the very least, draw your conclusions into question.

The Content/Noncontent Debate

While a forensic analyst could hope for more and better backing, cusum has definitely proven itself to be a useful tool. It is surprising, then, to find the technique, and its users, being attacked extremely viciously in some corners of academic research. Once again, this could be related to the distinction between content and noncontent analysis. Content-based examination is obvious and fairly intuitive. Noncontent methods can seem strange until the supporting studies are available and understood.

When dealing with authorship analysis of text, it may be important to distinguish between issues of literary style, and *stylometry*, the numeric measures such as those involved in noncontent analysis. Literary critics, and anyone with a writing background, may be prejudiced against technologies that ignore the content of the text and concentrate on other factors. Although techniques such as cusum analysis have been proven to work in practice, they still engender unreasoning opposition from many who fail to understand that text can contain features quite apart from the content and meaning.

It may seem strange to use meaningless features as evidence. However, Richard Forsyth reported on studies and experiments that found short substrings of letter sequences can be effective in identifying authors. Even a relative count of the use of single letters can be characteristic of authors. This latter result has definite implications in regard to software forensics. If analysis of the relative use of specific bytes, or even opcodes, can be used as an identification metric, forensic analysis of code with respect to authorship could be greatly assisted.

Forsyth's research has additional implications for software analysis. His original observation was that the metrics used in text examination often seemed arbitrary and yet provided impressive results. He then set out to have the analysis program itself search for factors that could be used as measures of author identity. As we shall see in Chapter 9 on authorship analysis, having a programmatic means of determining signatures that should be examined in code analysis would be a big step forward.

Noncontent Metrics as Evidence of Authorship

The kinds of measures that cusum and short string analysis can bring to forensic linguistics can be impressive, but they are definitely limited. Like physical fingerprints, these calculations can confirm or verify identification, but they generally cannot provide us with the identity of an author.

The situations where noncontent analysis can help us are usually defined within narrow parameters. If the number of possible authors is small, we can probably easily find some distinguishing characteristics. In a large potential population, we do not yet know how well we can determine which one is the author we want.

For each author that we wish to compare against our subject text, we need to have a body of work that is known as theirs in order to extract sample signature patterns. The known portfolio must be large enough that we can reliably derive characteristics that will identify the author. In the case of cusum, samples as small as 25 sentences seem to work well, but other methods may require more text.

We have noted that cusum is able to find characteristics that are consistent over time, but that may not be true of all noncontent signatures. In the case of variable patterns, the body of work used as a basis should span a long period of time, so that trends may be determined.

Noncontent metrics are not as contingent on variations due to subject or genre, as are content-based patterns. However, we should be aware of the possibility of such irregularities and ensure that our sample either spans a number of styles or is close to the passage under consideration.

Additional Indicators

Certain message formats may provide forensic linguists with additional information, although this is still primarily in the realm of traditional computer forensics. A number of Microsoft email systems include a data block with every message that is sent. To most readers, this block contains meaningless garbage. However, it may include a variety of information, such as part of the structure of the file system on the sender's machine, the sender's registered identity, programs in use, and so forth.

Other programs may add information that can be used. Microsoft's word processing program, Word, for example, is frequently used to create documents sent by email. Word documents include information about file system structure, the author's name (and possibly company), and a "global user ID." This ID was analyzed as evidence in the case of the Melissa virus. Microsoft Word can provide us with even more data: Comments and "deleted" sections of text may be retained in Word files and simply be marked as hidden to prevent them from being displayed. Simple utility tools, such as hex editors or even text editors that can read Microsoft Office file formats, can recover this information from the file itself.

Summary

The procedures of forensic linguistics and stylistic analysis can be applied, with some modification, directly to the analysis of source code. Object code does not have the structure of words, sentences, and paragraphs used in text. The concepts involved in linguistic analysis of authorship do have a relationship to techniques already directed toward the study of software, and therefore, indicate fruitful areas of research.

9

Authorship Analysis

In the preceding chapter on forensic linguistics, we noted that writers can be identified by a particular writing style. We can find the same kind of stylistic indicators in the work of a programmer, as well. In addition, we can find signatures that are not part of the content or structure of the program, which still give us information about the author. Over time, we will develop even more of these metrics, making identification more usable, accurate, and reliable.

Problems

One factor that must be addressed immediately is to determine the characteristics that distinguish a specific programmer from other coders. In looking at a program, we may notice very unique and distinctive signs, but these are useless, for purposes of identifying the author, if they don't hold true for the author overall. To return to an example from literature, when examining Ian Fleming's *The Spy Who Loved Me*, note that it was written in the first person, from the point of view of a subordinate character. This is an unusual, though hardly unique style. But when looking at the rest of Fleming's work, we find that it was unusual for him as well: Nearly all of his writing (the James Bond stuff, anyway) is written from the point of view of the universal narrator. So this stylistic indicator, from a single example, is unreliable as an identifier. For author identification, we need to find a set of characteristics that remain constant for a significant portion of the programs that an individual coder might produce.

Software evolves over time, when the author becomes dissatisfied with it or must accommodate forces in the marketplace. Programmers vary their programming habits and their choice of languages and tools. Software gets reused, both by the original programmer and sometimes by others. We must be careful to allow for all of these possibilities when conducting forensic analysis.

We expect the programming mannerisms of programmers to change and evolve. Education and other factors have an effect on the development of programming styles. Software engineering models impose naming conventions, parameter passing methods, and commenting styles, as well as a development strategy. (These influences from training and tools may become cultural indicators, and thus allow us to narrow our field of interest to a certain group.) As an example, the waterfall model encourages the design of precise specifications and the use of modules. This type of development has an impact on the style of the resulting program. In addition, the programming style varies from language to language, or because of superficial restrictions imposed by managers or environments.

It also seems unlikely that, given a piece of software, we will be able to identify the programmer who wrote it out of the millions of coders, both professional and amateur, who develop software. We must use software forensics in the same way that we use other forms of forensic evidence for identification. We may need to narrow our search using particular points of information, such as broad cultural indicators, before we can make a final determination about identity.

Plagiarism Detection versus Authorship Analysis

Although we may be using both, and the techniques of one can assist in the other, we need to make a distinction between authorship analysis and plagiarism detection. Plagiarism can be defined in terms of the replication of software (or other material), with or without the permission of the original author. Therefore, it may not always be necessary to determine whether the respective programs were written by the same person.

Plagiarism detection detects or measures similarity between two programs. Authorship analysis does not. In fact, in authorship analysis, we may need to be able to find similar styles or characteristics in two wildly differing programs.

In general, when dealing with plagiarism, we will be comparing apples to apples, rather than oranges. That is, to determine if code has been stolen or reused, we will compare machine code with machine code, C with C, Python with Python. We don't often have to determine characteristics that can be compared across languages. The same is not necessarily true with respect to authorship analysis. A programmer will typically know, and work with, a number of programming languages. A preference for a language may be one characteristic of a coder (and may show up even in analysis of the object code), but a programmer is usually far from limited to that one language and may learn others as required. Therefore, in authorship analysis, we may need to develop signatures and measures that we can compare even between programs written in different languages.

From this initial discussion, it may be felt that plagiarism detection is, by far, the simpler task. After all, you must only find the similarities between two programs, not find one programmer out of millions. In practice, this difference vanishes. A useful plagiarism detection system will have to obtain samples from a wide variety of sources. As noted in the earlier chapter on advanced tools, plagiarism detection systems in colleges may have a necessity to do Web searches to find roughly similar papers or code samples, and then perform exact matching. On the other hand, when called on to do authorship analysis, we may only need to confirm the identity of a questionable individual, or to find one out of a small population of suspects.

To clarify this point, we should also consider different types of activities that can all be considered as authorship analysis. In the first case, we may simply need to distinguish between the authors of different programs, or different sections of a program. We are not assigning names to the creators of the different bodies of work, but only determining that they are dissimilar. At a second level, we look for the attributes and idiosyncrasies of a coder. These characteristics may be the type of information discussed in Chapter 7 on programming cultures: educational, community, and social traits. Some signatures at this level may tell us personality traits and other idiosyncratic information about the author, but without specifically identifying him or her. It is only at the final stage that we attempt to determine a unique identity.

How Can It Work?

We are creatures of habit. When we go to the grocery store and walk through the aisles, we start on the same side, each time. In college, we sit in the same seat for an entire year. People work within certain repeated frameworks and patterns. We like things that are more comfortable or to which we are accustomed. We use the same tools (such as 15-year-old word processing programs) in the same way (some write macros to automate every repeated task; others can't be bothered) time after time.

Programming, although the specific tasks and functions of the software are different for each project, has many factors that can remain constant. Coders have preferences for certain data structures and algorithms, compilers and other utilities, and certain system calls or library functions. When examining source code, there are styles involved in formatting the code, and in the use and creation of variable and function names. There will also be evidence of the program author's level of expertise, in relation to the maturity, skill, and elegance that is manifest in the code.

Programmers tend to repeat patterns. Coders are familiar with a circumscribed set of paradigms, mainly those they know well and use frequently. Intimacy with these exemplars helps them to write programs faster and more reliably. It would be unrealistic to assume that any programmer could develop programs efficiently and correctly using an unfamiliar programming style. This need for familiarity applies to the structure of the programs and also to the display and layout of the source code on the screen, so that the logical structure of the software makes sense to the author.

Source Code Indicators

When examining source code, we can observe textual or typographic characteristics. Some of these signatures may be:

- Comments on the same line as code statements and commands.
- Multiple comments occurring together in a block.
- Comments set off by a border such as repeating characters.
- The amount of indentation distinguishing a function block from other code around it.
- Uppercase or lowercase characters used exclusively.

- Case used to distinguish between different types of code entities, such as commands and variables.
- The keyword for the beginning of a block followed by a command statement on the same line.
- Multiple statements per line.
- The amount of "whitespace," including indentation and blank lines in the program body.

(Note: In regard to this list of factors, some of them may be specific to a language. For example, not all programming languages allow a comment to be on the same line as a command or statement.)

Additional points in source code may be of great assistance to us. One factor involves errors in writing. Many programmers have difficulty writing correct prose. Variable names are frequently misspelled, and mistakes inside comments may be quite telling if the misspelling is consistent. (Variable and function names may also be prone to spelling errors, but must be consistent or they will be flagged as errors by compilers.) Grammatical mistakes inside comments or print statements might provide an additional point of similarity between two programs.

More General Indicators

In general, typographical signatures, such as those that rely on the amount of whitespace provided in the source code, can be strongly influenced by automated utilities that help to keep source code clean and legible. Therefore, these will be less useful as indicators of authorship. More reliable are those dealing with items completely under the control of the programmer, such as the choice (and even length) of variable names. However, those items may not survive the compilation into source code. Most reliable are those factors that are fundamental to the program structure. These could include the type of control loops preferred by the author, the tendency to check (and allow) for error conditions, and the ability of a module to "do nothing" without creating a problem. Proper error checking is unfortunately uncommon in programming, and therefore, the existence of such code is evidence of a seasoned programmer. In addition, purely because of the infrequency of exception handling, authors may develop characteristic ways to deal with it.

Other factors that survive compilation deal with the choice of data structures and algorithms. In the investigations into the Internet Worm of 1988, for example, it was found that the program built large lists and repeatedly searched them. This procedure reduced the performance of the program: The searching could have been conducted more quickly with other data structures and search algorithms, such as those related to hashed indices. The author had been trained in conceptual, rather than practical methods, and was not concerned with optimizing the performance of his creation. He had learned programming in the LISP computer language, which promotes the use of linked lists as data structures. (The name LISP comes from a contraction of LISt Processing.)

We can judge programming experience almost by measuring laziness. Coders do not like to reinvent the wheel. A novice will probably not know all of the shortcuts that are available, nor the help that the system can offer. A code maven will use functions that the system provides. An experienced programmer will also make use of recursion, an abstract and difficult construct that often shortens source code considerably.

There are also signs of inexperience. Code that is never called in the program may indicate carelessness in failing to remove modules that have been replaced by later versions. Other unused code may relate to debugging functions or features that the author neglected to elide from the final version.

Is It Reliable?

In academic testing, this hypothesis has been assessed against small samples of programs and authors. The results were encouragingly accurate. More testing does need to be done to determine how useful this technique can be in a broad range of environments and with a larger population of programmers.

Some find it puzzling that even quickly chosen classification metrics are able to identify a programmer, even when a coding style was deliberately varied. It is likely that people are more consistent in their choices than they realize, in directions that are not evident to the casual observer.

Summary

While it may seem particularly odd that we should actually be able to find clues to the identity of a programmer in the software he or she writes, this chapter has briefly pointed out a number of characteristics that can be assessed simply from the code itself. Both source and object code can provide indications about the background, personality, training, and general practices of an author. Of course, if the author also happens to embed text strings to include his or her name, address, and Web site URL, that makes our job even easier.

References and Resources

This book is a mere overview of the multiple fields and technologies related to software forensics. While it brings together the basic concepts, there is a great deal of material in each area that cannot be addressed in a single book. The following are resources for further work and exploration if you wish to pursue more detailed study.

As noted in the Preface, this section concentrates on books and online sources that are most available to professionals outside the academic world. You may wish to look more deeply into the academic research for specific points.

An excellent source of research papers is the NEC Research Institute ResearchIndex (also known as CiteSeer), at http://citeseer.nj.nec.com/. The citation index allows you to quickly find papers on the same or related topics.

For example, a paper reporting on work to extend literary authorship analysis to source code is at http://citeseer.nj.nec.com/gray97software.html, and another on authorship analysis that examines the use of fuzzy logic can be found at http://citeseer.nj.nec.com/kilgour97fuzzy.html. Not only is the full text of these papers available, but by following links, you can find research the authors used themselves, and other essays that cited their work, resulting in quick access to research on a whole range of related topics.

Papers related to security and software forensics may also be found at the COAST Library, http://www.cerias.purdue.edu/

coast/coast-library.html, such as an early paper on tracking code to its authors: http://www.cerias.purdue.edu/homes/spaf/tech-reps/9210.ps.

A number of academic papers are available online in portable document format (.PDF) or PostScript (.PS). These two formats may be viewed with the GhostView add-on for Ghostscript, and both programs are freely available. Information on downloading the software may be found at the Ghostscript home page (http://www.cs.wisc.edu/~ghost/), Ghostscript.com (http://www.ghostscript.com/), GNU Project Ghostscript (http://www.gnu.org/software/ghostscript/ghostscript.html), and GNU Project Ghostview (http://www.gnu.org/software/ghostview/ghostview.html).

Introduction and Background

Computer forensics has traditionally been seen primarily in terms of the recovery and preservation for presentation as evidence of data from computers that may have been used in the commission of some criminal activity. As such, most of the existing books on computer forensics do not deal directly with software forensics. The recovery of software, though, is the same as the recovery of any information from a computer. Therefore, traditional computer forensics works present guides to follow in the recovery and preservation of software for evidence.

Computer Forensics: Incident Response Essentials
Warren G. Kruse II and Jay G. Heiser
Publisher: Addison-Wesley Publishing Co.
P.O. Box 520, 26 Prince Andrew Place
Don Mills, Ontario M3C 2T8
Year: 2002
ISBN: 0-201-70719-5

The text by Kruse and Heiser is probably the best single work in the field. I'm still disappointed that authors seem to think computer forensics is limited to data recovery, but this work at least has utility value going for it.

Chapter 1 is a rough outline of data recovery, with an emphasis on documentation and the chain of evidence. It is heartening to see the inclusion of network forensics, in that the book also contains basic information about IP addressing for the purpose of tracing network intruders: It is useful and does not drown the reader in inconsequential details. A valuable discussion of email headers, as well as a very terse outline of intrusion detection systems (IDS) is also included. The material on hard drive basics and concepts is generally good, but some points on imaging and connecting are passed over rather quickly. Chapter 4 has a reasonable high-level overview of encryption abstractions, but it is difficult to see the immediate relevance of the material to forensics. The description of hostile code, in Chapter 6, matches that of weeds in gardening: anything you don't want. It is, from a software forensics perspective, disappointing but unsurprising to find that the content, while basically sound, is not particularly structured or helpful.

Software (and some hardware) tools are described, and the text explains a number of points about the Windows operating system that might affect data recovery and forensics. The introduction to UNIX is more structured and detailed, although it examines fewer specific tools. A wide variety of tools and commands for collecting information from and about UNIX systems is given briefly in Chapter 11. In addition, there is a short introduction to general concepts in the (U.S.) law enforcement system.

Computer forensics books are starting to come out of the woodwork, and most offer such sage advice as "gather evidence" and "don't mess up the chain of custody." This book does tend to follow the same style and tone, but also has very valuable tips for practical work. It won't help you much in analysis, particularly of software, but it will help you become better at collecting data that will stand up in court.

Computer Forensics and Privacy
Michael A. Caloyannides
Publisher: Artech House/Horizon
685 Canton Street, Norwood, MA 02062
Year: 2001
ISBN: 1-58053-283-7

This book occupies a unique place in the literature of computer forensics. Most works in the field, such as Kruse and Heiser's *Computer Forensics* (above), concentrate on documentation of the investigation, with a view to presentation in court. The actual mechanics of data recovery tend to be left to commercial tools. Caloyannides demonstrates how to delve into corners of the computer to actually get the data out.

At the same time, this work is inconsistent on at least two levels. The perspective flips back and forth between forensics and privacy, alternately emphasizing how to find evidence and how to hide evidence. The technology involved is the same, but the shifts in viewpoint can be jarring to the reader. In addition, the depth of technical detail can vary wildly. At one point, the book stops shy of telling you how to undelete files with a sector editor (an activity that could be useful to every computer user), while other sections list lengthy and extraordinary measures to secure personal computers.

Part one concentrates on the data recovery aspect of computer forensics. The use of, and factors related to the use of computer forensics is supported by specific cases (rather than vague suppositions). The book outlines various places (primarily in Windows) from which data may be recovered. It is an odd mix of little known and very valuable information, and extremely poor explanations of basic functions like manual undeletion and file overwriting. There is also a strange and terse look at steganography, U.S. and U.K. surveillance systems, cryptography, and anonymity. Data acquisition, from sources such as key logging and Van Eck radiation, is reviewed, and a short list of measures falsely believed to provide privacy protection is debunked.

Part two turns to privacy and security. Chapter 8 is a discussion of legal and commercial protections of privacy (mostly in the US) and their failings. Installing and tweaking a privacy-protected configuration of Windows is covered in considerable detail. Intermediate online privacy looks at browser and email configurations and security packages and has a section on tracing email that would be helpful in dealing with spam. The more advanced section, dealing with anonymizing services and personal firewalls, may be beyond the average user. A general opinion piece on cryptography nevertheless provides a good, basic background, albeit with a social and political emphasis. Chapter 14 looks at more practical encryption, detailing PGP and specialized cryptographic programs, with a detour into biometrics. Part three is a brief look at legal and other issues.

Despite the ragged organization and style, and some glaring gaps in coverage, this book does contain a wealth of information for both the computer forensic examiner and the user concerned with privacy. For anyone beyond the most basic user, it is well worth a read.

Desktop Witness: The Do's and Don'ts of Personal Computer Security
Michael A. Caloyannides
Publisher: John Wiley & Sons, Inc.
5353 Dundas Street West, 4th Floor, Etobicoke, Ontario M9B 6H8
Year: 2002
ISBN: 0-471-48657-4

Caloyannides' more recent work takes very much the same approach, but does contain different, and also useful information in regard to computer forensics.

The title and the subtitle of this book are somewhat at odds. Is this text about the evidence that can be extracted from desktop machines? Or is it about protecting yourself and your personal computer or information? Caloyannides seems to be making the point that the answer is both: that there is an overwhelming need to ensure that your computer isn't finking on you, and that you must make every effort to ensure that the government cannot obtain the information on your desktop. While he is clearly on the personal side of the privacy versus national security debate, even those who agree with him may find the arguments shrill and extreme. Caloyannides may have hurt his own case by taking an anarchistic and almost paranoid position in stating the need for privacy against government encroachment.

The book lists a vast array of measures that may protect privacy. Some are questionable: Caloyannides makes a blanket recommendation to install all operating system patches, but notes that doing so for some versions of Windows requires you to give away a lot of information. He does not, though, detail the times that official patches have made the situation worse rather than better, nor the complexity of some patches. By mid-2002, one expert noted that an effective installation of the Windows NT operating system required 29 steps, including no less than three separate installations of the latest service pack at different points.

While there are issues of general security in the book, it is, first and last, about privacy, and primarily personal privacy. The material could have been structured more usefully and written less stridently, but a great deal of helpful content is included. Those interested in privacy will find it interesting, and computer forensic specialists may also find it to be a handy reference.

Computer and Intrusion Forensics
George Mohay, Alison Anderson, Byron Collie, Olivier de Vel, and Rodney McKemmish
Publisher: Artech House/Horizon
685 Canton Street, Norwood, MA 02062
Year: 2003
ISBN: 1-58053-369-8

Computer and Intrusion Forensics is the first real attempt to bring both computer and network forensic topics into a single book. (It is intriguing to note that Eugene Spafford, who wrote the Foreword, is a pioneer of the "third leg" of software forensics. The book does not cover it at all.)

An introduction to computer and network (intrusion) forensics, points out the ways that computers can be involved in the commission of crimes, and the requirements for obtaining and preserving evidence in such cases, as well as a broad overview of the concerns, technologies, applications, procedures, and legislation bearing on digital evidence recovery from computers. In fact, Chapter 2 is the equivalent of, and sometimes superior to a number of the computer forensics books available. However, the breadth of the discussion does come at the expense of depth. This content is quite suitable for the information security or even legal professional who needs to understand the field of computer forensics, but it does not have the detail that a practitioner may require. Forensic accounting and the algorithms that can be used to detect fraud are outlined, but very little is directly relevant to computer forensics as such.

While the computer forensic content is sound, and it is heartening to see other fields being included, the very limited work on network forensics is disappointing. This text is a useful reference for those needing background material on forensic technologies, but breaks no new ground.

To date, network forensics has not been addressed as such in a text. The most relevant works on the topic are those dealing with intrusion detection. At least four books are identically titled *Intrusion Detection*. The following two should not be confused with the others.

Intrusion Detection
Edward G. Amoroso
Publisher: Intrusion.Net Books
P.O. Box 78, Sparta, NJ 07871
Year: 1999
ISNB: 0-9666700-7-8

Amoroso has particularly chosen to avoid specific software products, concentrating on concepts and technical details. The text is based on material for an advanced course in intrusion detection, but is intended for administrators and system designers with a security job to do.

After demonstrating that the intrusion detection term means different things to different people, the author gives us an excellent, practical, real-world definition. This is used as the basis for an examination of essential components and issues to be dealt with as the book proceeds. Five different processes for detecting intrusions are discussed. Each method spawns a number of "case studies," which, for Amoroso, means looking at how specific tools can be used. (This style is far more useful than the normal business case studies that are long on who did what and very short on how.) Intrusion detection architecture is reviewed, enlarging the conceptual model to produce an overall system. Chapter 4 defines intrusions in a way that may seem strange, until you realize that it is a very functional description for building detection rules. The problem of determining identity on a TCP/IP internetwork is discussed, but while the topic is relevant to intrusion detection, few answers are presented.

The bibliography is, for once, annotated. Books and online texts are included, although the emphasis is on journal articles and conference papers.

The content is readable and, although it seems odd to use the word in relation to a security work, even fun. The book works on a great many levels. It provides an overall framework for thinking about security. It thoroughly explains the concepts behind intrusion detection. And it gives some very practical and useful advice for system protection for a variety of operating systems and using a number of tools.

Intrusion Detection
Rebecca Gurley Bace
Publisher: Macmillan Computer Publishing (MCP)
201 W. 103rd Street, Indianapolis, IN 46290
Year: 2000
ISBN: 1-57870-185-6

Bace's take on this topic (and title) provides a solid and comprehensive background for anyone pursuing the subject. Concentrating on a conceptual model, the book is occasionally weak in regard to practical implementation, but more than makes up for this textual deficiency with a strong sense of historical background, developmental approaches, and references to specific implementations that the practitioner may research separately.

Chapter 1 presents a history of intrusion detection, starting with system accounting, through audit systems, to the most recent research and experimental systems. The definitions and concepts focus from broad security theory to specific intrusion detection principles and variants. Intrusion detection requires analysis of system and other information, and the book describes the sources for this data. The review of possible responses includes warnings against inappropriate over-reactions. Vulnerability analysis, including a close look at controversial tools like COPS, SATAN, and ISS is covered.

The work talks about technical issues that are still to be addressed, and about legal issues, evidence, and privacy. Security administrators and strategists, at the executive level, are presented with everything from the need for security goals to globalization. Designers get a few general guidelines, along with comments from those who have been implementing exemplary systems. There is a realistic look at future developments in attacks and defense.

Intrusion Signatures and Analysis
Stephen Northcutt, Mark Cooper, Matt Fearnow, and Karen Frederick
Publisher: Macmillan Computer Publishing (MCP)
201 W. 103rd Street, Indianapolis, IN 46290
Year: 2001
ISBN: 0-7357-1063-5

In concept, this is a very badly needed book. Intrusion detection and network forensics are now vitally important topics in the security arena. An explanation of how to identify dangerous signatures and extract evidence of an intrusion or attack from network logs is something that most network administrators really need. Unfortunately, while the idea is good, the execution, in the case of the current work, is seriously flawed.

There is a good deal of valuable material in this book. Unfortunately, it is not easy to extract the useful bits. The new system administrator will not find the explanations clear or illuminating. The experienced professional will not find particular attacks or traffic types easy to find for reference. Both groups will find themselves flipping back and forth between sections of the book, or even between sections of the exegesis of one particular attack. However, both groups will likely be interested in the book anyway, simply because of the lack of other sources.

The following rather odd book could be considered under network forensics.

Hacker's Challenge
Mike Schiffman
Publisher: McGraw-Hill Ryerson/Osborne
300 Water Street, Whitby, Ontario L1N 9B6
Year: 2001
ISBN: 0-07-219384-0

Initially, I was skeptical of the title, considering the wording to be simply jumping on the current security bandwagon, with "hacker" this and "hacker" that on every bookshelf. In an odd way, however, the title is quite appropriate. This volume contains a series of 20 cases (or tests) that are supposed to challenge your ability to analyze network data (most of the scenarios are network based) to identify and assess intrusions. Unfortunately, there are some problems in the implementation.

The book is divided into two parts. First come the 20 scenarios, with varying types and degrees of detail about the problems. Then come 20 "solutions," which are supposed to point out how you should have approached the situation, and what indicators should have tipped you off to the intrusion and intruder. This physical division is rather meaningless. It isn't as if the solutions were short phrases that had to be printed upside down at the bottom of the page so that the reader doesn't inadvertently read the answer to the riddle while thinking about it. There is no reason that the solutions could not immediately follow the stories.

Actually, the pieces were written by 13 different authors, and the amount of detail varies tremendously. Therefore, all the possible mistakes that could be made in a work of this type are represented. Sometimes the audit logs presented in the scenario contain the relevant details and very little else, with very sparse explanations. In other pieces, readers are presented with huge amounts of log data and the relevant points are lost. Some scenarios are not complete, and the data necessary to solve the problem are not given until the solution write-up. A few pieces contain almost no data for the reader in the problem section, while the solution presents almost no detection information or forensic exegesis. One case gives pages of log data and almost no analysis at all in the solution. Some articles simply reproduce earlier situations with different characters. One solution makes no sense in terms of the data given in the problem outline. Some pieces are unclear, some simplistic, and some can only be described as misleading.

The occasional scenario is written up almost poetically, and isolated solutions do have tutelary explanations of how to read network audit logs.

If you are very good at forensic network analysis, you might enjoy pitting yourself against these challenges.

Email analysis is a rather specialized field, possibly slightly removed from network forensics itself. To date, most of the literature in that field has concentrated on the phenomenon of unsolicited commercial email, otherwise known as spam.

Removing the Spam: Email Processing and Filtering
Geoff Mulligan
Publisher:Addison-Wesley Publishing Co.
P.O. Box 520, 26 Prince Andrew Place
Don Mills, Ontario M3C 2T8
Year: 1999
ISBN: 0-201-37957-0

This book is intended for the system manager, rather than the end user. More specifically, it is aimed at the mail administrator for an Internet service provider (ISP) or corporate network. Slightly unfortunate is the fact that it becomes more particular still, being of greatest use to those running UNIX, sendmail, Procmail, and either Majordomo or SmartList.

Chapter 1 is an excellent overview of electronic mail. It is concise, complete, and accurate. Newcomers to the field will find not only a conceptual foundation for all the aspects of Internet email, but also pointers to other references. Professionals will find fast access to a number of details that need to be addressed on a fairly frequent basis. The main theme, of course, is how spam uses the functions of email systems, and how it can be impeded, with as little impact as possible on normal communications. A good framework is presented in this chapter, with a number of references to spam-fighting resources. If I were to make one suggestion, it would be to increase the number of examples of forged email headers and how to dissect them.

Stopping Spam
Alan Schwartz and Simson Garfinkel
Publisher: O'Reilly & Associates, Inc.
103 Morris Street, Suite A, Sebastopol, CA 95472
Year: 1998
ISBN: 1-56592-388-X

Eternal vigilance is the price of junk free email. Therefore, readers expecting to find a quick fix for spam in this book are possibly going to be disappointed. Those who persevere, however, will find much useful material that is both interesting and valuable in the fight against unsolicited and commercial mass mail bombing.

The book provides a solid technical background for further discussion of spam, covering mail agents and the mail and news protocols. A number of steps that the average computer user can take are listed.

The work provides a good deal of helpful and useful information for Internet service providers (ISPs), corporate network administrators, and net help desks.

A number of other books on computer forensics and incident response are reviewed on the Web pages at http://victoria.tc.ca/techrev/mnbkscli.htm or http://sun.soci.niu.edu/~rslade/mnbkscli.htm.

In terms of resources on the Web, some may be found at the Digital Forensic Research Workshop at http://www.dfrws.org/. A relatively recent and (currently) low traffic mailing list on computer forensics is available at http://groups.yahoo.com/group/computerforensicsworld/.

In the network forensics arena, an early paper is Anderson's *Computer Security Threat Monitoring and Surveillance* at http://csrc.nist.gov/publications/history/ande80.pdf. Peter Neumann has been instrumental in developing the EMERALD model for SRI, and early work with Dorothy Denning on *An Intrusion Detection Model* is reported at www.cs.georgetown.edu/~denning/infosec/ids-model.rtf. (This paper also has implications for software anomaly analysis.)

166 Blackhats

The general public is usually fascinated by the "dark side" of computing and the mysterious activities that go on there. There are some good reasons why it is necessary to study what are frequently, and mistakenly, referred to as "hacker" communities. The first being that "blackhats" are almost universally seen as "evil geniuses," and this characterization is wrong on at least two counts. The more important reason is that examination of the various types of blackhat communities, along with their activities and motivations, provides valuable direction for the potential success of the forensic programming endeavor, as well as cultural indicators and signatures that may be of use in the process.

Because of the extensive availability of poorly researched material in this field, all references should be examined carefully. Probably the best qualitative and ethnographic research has been done by Sarah Gordon. *The Generic Virus Writer* papers make heavy use of interviews with a handful of virus writers and challenge all the stereotypes. Many of Gordon's papers are available at http://www.badguys.org/papers.htm. Dorothy Denning has also done some serious work in this regard, such as her article in the book *Computers Under Attack*.

Although it has problems, probably the best text in this area is the following:

Hackers: Crime in the Digital Sublime
Paul A. Taylor
Publisher: Routledge
11 New Fetter Lane, London, England EC4P 4EE
Year: 1999
ISBN: 0-415-18072-4

Following in the footsteps of Gordon, Denning, and Ray Kaplan, Paul Taylor is attempting to open the world, and world view, of those who make informal attempts to penetrate computer and communications security to the security "expert." The book tries to explain motivations, culture, and background, with a view to the benefits of a dialog between the official guardians and those who pry at the gaps in the armor. Using extensive interviews with people from both sides of the divide, Taylor attempts to put forward the reality behind the hype.

The book examines the terminology—hack, hacker, and hacking—emphasizing the original meaning of creative and useful mastery of the technology. Hacking culture is reviewed quite thoroughly, although perhaps not enough attention is paid to the divisions and continuum that exists. (I was amused by the note in the Preface to the effect that nobody would admit to distributing viruses. Virus writers still occupy the lowest rung of the blackhat ladder.) Motivation is explored, though possibly too much credence is given to self-reporting. Chapter 4 is a marvel, a first-rate examination, and an indictment of the state of computer security (or, perhaps, insecurity). Arguments for and against, as well as dialog with and employment of those who have done unauthorized security breaking are explored as well. Chapter 6, however, turns to presenting a number of sociological theories about why hackers might be marginalized. This material seems to have no purpose other than to propose that such people are being treated unfairly. It is disappointing that the discussion of ethics presents little content that is germane to the discussion and seems to wander off into miscellaneous speculation.

Taylor is making a conscious effort to avoid sensationalism, and, indeed, to counter the sensational and misinformed reports of computer security penetration that are prevalent in the popular media. His writing style is more "academic" than is necessary, using, for example, the passive voice most of the time. (Oddly, for all its academic formality, endnotes, and bibliography, the work falls short in terms of clarity of references and citations.)

The extensive use of interview materials and quotations from other works are both a strength and a weakness. No one perspective is allowed to dominate, and a great many arguments and opinions are presented. The constant quotes from a variety of sources, however, often reduce the readability of the work. I found the book very difficult and time consuming to get through. Added to this, Taylor's aversion to contaminating the source material with his own analysis ensures that the text is very demanding of the reader's own analytical skills and work.

Taylor does make a serious effort to give a fair and even presentation to both sides of the argument, but it is still fairly obvious that his sympathies lie in "detente." The title of the book itself indicates this. There is a discussion of the derivation and evolution of the "hacker" term, but the acceptance of the "popular" status of the word to mean those who break into computers also allows those who break into computer systems to present arguments for their behavior as a kind of discovery learning, without the supporting evidence that would otherwise be necessary. In this, Taylor's work shares a weakness with other similar books on the topic. "Hacker" claims are taken at their own valuation without much analysis of either factual or motivational claims.

Even with the problems presented above, I still highly recommend this work to anyone in the security field, or to anyone who wants to understand either security work or an important part of the computer culture. For all its flaws, Taylor's book is the most extensive and detailed examination of the cracker phenomenon I have ever read.

Although it may seem odd, a very good study of the culture that blackhats want to emulate is in the following:

The New Hacker's Dictionary
Eric S. Raymond
Publisher: MIT Press
55 Hayward Street, Cambridge, MA 02142-1399
Year: 1996
ISBN: 0-262-68092-0

That the book is a source of amusement and entertainment is undeniable. Raymond and company have, however, compiled substantial material of social, cultural, and historic value for those wishing to understand both the strict hacker culture, and the more diffused genre of technical enthusiasts that surrounds computing and computer networks.

The linguistic analysis of hacker culture is a scholarly work in itself. Whether linguists accept it as such in their own field, this work has done the field work and compilation for them. The analysis is incisive: I was quite startled to find the undoubted source for my own discomfort with including punctuation inside of quotation marks.

You can, of course, find any number of books purporting to explore and explain the "dark side" community, some of which are little more than fiction. While Taylor's book has fundamental flaws, he has a great deal more material and a wider range of direct contacts than Levy, Sterling, Dreyfus, or Verton. His conclusions are significantly more reliable, but you may wish to examine some of the other works.

Hackers
Steven Levy
Publisher: Dell Publishing
1540 Broadway, New York, NY 10036
Year: 1984
ISBN: 0-385-31210-5

The title covers more than the book actually examines. The hacker culture is not viewed as a whole, but only as isolated, though admittedly important pockets. The first part of the book deals with the early days of the MIT lab, the second with the Homebrew Computer Club and the growth of the microcomputer industry. Even within these limited confines, the book concentrates on specific individuals within the groups, rather than communities as a whole. The third part is even more limited, dealing with a single game company and basically two principals.

The material gathered and presented here is a lot of fun and contains a great deal of anecdotal information. Not a work of scholarship, the book is vastly entertaining. The stories, following little logic or thread, do flow almost seamlessly, one into another: Levy's writing style is very readable. The computer folklore–literate will find not only the known, but esoteric bits of trivia, as well. Those who have never had anything to do with computers need not fear being left out because of technical detail: There isn't any. The stories of these technical wizards and objects deal exclusively with the human side. Levy seems to select for the bizarre and the useless. The personalities and projects presented are the odd or the playful. Those, indeed, who do have a concept of utility or service receive little sympathy in the book. The reader, without other sources, will undoubtedly fix on the idea of hackers as ephemeral game players of no value whatsoever.

The Hacker Crackdown
Bruce Sterling
Publisher: Bantam Books
1540 Broadway, New York, NY 10036
Year: 1992
ISBN: 0-553-56370-X

It is important to keep in mind that the crackdown of the title refers to a specific incident: the series of raids in 1990 by various U. S. law enforcement agencies that tend to be collectively, if incorrectly, subsumed under the code name "Operation Sundevil." The book brings together a number of the stories surrounding this event, as well as gives some background, particularly in regard to AT&T and the U.S. Secret Service. There are, however, significant gaps that prevent it from being an overall analysis of either the cracker/phone phreak culture or the data security/law enforcement community.

As an overview of the 1990 raids, the book is entertaining, often informative, and generally well written. Digressions often provide very interesting background, although at times they consume entire chapters without much bearing on the central issues. Those who were around for the electronic discussions of the 1990 raids will possibly be glad of the collection of all the stories into one place. (Those who have dealt with blackhats will readily recognize some of the descriptions, as well as the repeated emphasis on braggadocio as a primary character trait.)

Although Sterling is aware of the debate over the term "hacker"—indeed, he worries over contributing to the degradation of the term—he does not distinguish between the various communities of electronic outlaws. In fact, he states at one point that all are the same. Similarly, his contacts with law enforcement and data security people are limited. For these reasons, the book is not useful as a general introduction to the field.

The writing is highly opinionated. The U.S.-centric view of technology borders on jingoism. In general, neither law enforcement nor the cracking communities are seen with any favor. Although we can sympathize with Sterling's motivation in wanting to bring to light the injustice done to a friend, the extreme sarcasm that cloaks most of the first half of the book makes it difficult to understand what point he is trying to make.

Underground
Suelette Dreyfus
Publisher: Reed Books/Mandarin/Random House Australia
35 Cotham Road, Kew 3101, Australia
Year: 1997
ISBN: 1-86330-595-5

This book is yet another gee-whiz look at teenage mutant wannabe-high-tech-bandits. The stories revolve around a number of individuals with loose links to one particular bulletin board in Melbourne, Australia, all engaged in system intrusions and phone phreaking.

An immediate annoyance is the insistence of the author in referring to system breaking as "hacking." ("Cracking" seems to be reserved for breaking copy protection on games and other commercial software.) If any actual hacking takes place—creative or otherwise sophisticated use of the technology—it isn't apparent in the book. The descriptions of activities are vague, but generally appear to be simple "cookbook" uses of known security loopholes. This may not accurately reflect the events as they transpired because the author also betrays no depth of technical knowledge and seems to be willing to accept boasting as fact. The bibliography is impressively long until you realize that a number of the articles are never used or referenced in the text—at which point, you wonder how much material has even been read.

The structure and organization of the book is abrupt and sometimes difficult. Social or psychological observations are arbitrarily plunked into the middle of descriptions of system exploration, and, even though the paucity of dates makes it difficult to be sure, they don't appear to be in any chronological sequence, either. Those who have studied in the security field will recognize some names and even "handles," but the conceit of using only handles for members of the "underground" makes it difficult to know how much of the material to trust.

The book's attitude is also oddly inconsistent. In places, the crackers and phreaks are lauded as brilliant, antiestablishment heroes; but, by and large, they are portrayed as unsocialized, paranoid, spineless nonentities, who have no life skills beyond a few pieces of pseudotechnical knowledge used for playing vicious pranks. So thorough is this characterization that it comes as a total shock to find, in the Afterword, that not only do these people survive their court convictions, but also become important contributing members of society.

The author seems to feel quite free to point fingers in all directions. The absurdity of giving "look-see" intruders larger prison sentences than thieves or spies is pointed out, but not the difficulty of legally proving intent. After repeatedly hinting at police incompetence, brutality, and even corruption, the book ends with a rather weak statement implying that the situation is getting better. The common cracker assertion that if "sysadmins "don't want intruders, then they should secure their systems better, is not followed up with any discussion of, for example, surveys showing only one full-time security person per 5,000 employees, and only passing mention, by one of the ex-intruders, of the extreme difficulty in doing so. Poor family situations are used so frequently to justify illegal activities that one feels the need to point out that most products of "broken" homes do not become obsessive, paranoid loner criminals!

It is interesting to see a book written about a non-U.S. scene, and from a non-American perspective. Technically and journalistically, however, it has numerous problems.

The Hacker Diaries: Confessions of Teenage Hackers
Dan Verton
Publisher: McGraw-Hill Ryerson/Osborne
300 Water Street, Whitby, Ontario L1N 9B6
Year: 2002
ISBN: 0-07-222364-2

Teenage hackers are misunderstood. Definitions are for lamers; morality is a "bogus" concept. These noble idealists are questers after the Holy Grail of knowledge: problem solvers who are attempting to enlighten the masses. Given a little dedication, you too can, inside of six months, go from being a technopeasant to "knowing everything there [is] to know" about computers. Thus it is written in the Gospel of Verton.

Even if you ignore questions about the definition of what "hacking" actually is, and even if you leave aside the author's biased sympathy for rebels-without-a-clue, the introduction alone points out that Verton has not performed the research one would think minimal to such a project: reading the "popular" literature on the subject, never mind the more serious analyses by researchers like Denning and Gordon. How else can he make the statement that this book is the first ever to try and penetrate the veil of secrecy surrounding the computer vandal community, an assertion that must come as a bit of a shock to the authors of the books we've just discussed. It is, therefore, no surprise that this author gets basic factual information wrong, such as the confusion of the infamous Operation Sundevil with more successful prosecutions of computer crime.

Verton decries the blind and ignorant stereotyping of loners who are more comfortable with computers than with their peers, but he is, himself, guilty of promoting the same kind of confusion. The group targeted after the Columbine shootings was not the computer community but the Goths, who share almost no characteristics with hackers except for a slightly obsessive interest in an esoteric topic and a position outside the mainstream. (Well, possibly also an aversion to sunlight....) Verton has attempted to include "representative" examples of both maladjusted criminals and ethical hackers, but draws no distinctions between them and, indeed, seems to be trying to lump them all together.

Verton's writing seems clear and readable until you start to think about it. A story will say that A happened, then B happened, then C happened, then B happened, then D happened, and then B happened. Times are quite indefinite, but because the narrative is unclear even about simple sequences, it is not any real shock to find out that the author is not familiar with larger items of technical history, such as that UNIX predates VMS. Likewise, Verton isn't interested in having consistency get in the way of a good story, even if the story doesn't make any sense. Directions and motivations change suddenly and without apparent reason: Reading between the lines indicates that there is a lot that we aren't being told. Probably the author wasn't told, either. It sounds like he didn't even ask. (The interview subjects seem to have realized that they were dealing with a credulous author: Verton retails stories out of common urban legends and jokes without seeming to have identified them as such. Despite his credentials as a reporter for a computer trade magazine, Verton's technical knowledge is questionable—he doesn't know a denial of service attack from a reformat, nor that the Macintosh doesn't have a Windows Registry.)

If you want to start to examine this culture for yourself, the tamest but most legible, and certainly most commercial point of entry is *2600* magazine (http://www.2600.com/). Another "dark side" publication is *Phrack* (http://phrack.org/).

Tools

To perform forensic programming you need to be really good at assembly language or machine language programming and disassembly. To understand software forensics, you must understand machine language programming concepts, which is not quite as difficult.

The following is an excellent text explaining the concepts of assembly language programming, and some of the architecture of Intel processors, DOS, and Linux:

Assembly Language Step-by-Step, Second Edition
Jeff Duntemann
Publisher: John Wiley & Sons, Inc.
5353 Dundas Street West, 4th Floor, Etobicoke, Ontario M9B 6H8
Year: 2000
ISBN: 0-471-37523-3

Chapter 1 is an excellent explanation of what programming (especially low-level programming) is, by analogy to a "to do" list and a board game. Numbering and arithmetic in binary, octal, and hexadecimal is thoroughly demonstrated (with added practice!), and basic computer architecture is addressed. The pointers to emulators of old style computers may be useful, as well as interesting: It is much easier to program in machine language on the old kit computers than it is on modern machines with layers of interfaces.

The commands and functions of the NASM-IDE development environment and editor (provided on the CD-ROM) are listed and explained. DOS program file structures are explained somewhat vaguely. DEBUG, and its various operations, is put through its paces with some simple opcodes.

Chapter 8 lists a simple assembly language program and explains the various parts. Procedures, libraries, opcodes, and commands are covered.

The book moves from DOS into Linux, and covers the programming tools most useful in that operating system. Differences in system calls and the assembler format lead to sample code that works in the Linux system. A final chapter points at resources for further explorations and work in assembly programming.

This complete and detailed work does take the novice, with no previous programming assumed, through the basics to the point that the reader could start the process of discovery. It is readable (and funny enough to keep you going through the dry parts), provides all the necessary bits (sorry) including software, and is an excellent introduction for anyone wanting to find out what programming "down to the metal" is all about.

The author also has a Web page (http://www.duntemann.com/assembly.htm) with some recommendations for other references, as well as additional resources.

If you are serious about software forensics, you will need to be an expert not only in machine language programming (for your particular machine), but also in the internals of the operating system you study. You will need masses of references, many of which will not be available in book form. For the Wintel world alone, you should consider a number of the following texts.

Dissecting DOS
Michael Podanoffsky
Publisher: Addison-Wesley Publishing Company
1 Jacob Way, Reading, MA 01867-9984
Year: 1995
ISBN: 0-201-62687-X

As the subtitle says, this is a code-level look at the DOS operating system. This requires extensive citation of source code and, were it to be MS-, PC-, or DR-DOS to be displayed, Microsoft, IBM, or Novell might take exception to it. In this instance, Podanoffsky presents his own RxDOS system, and explains how MS-DOS differs.

Were RxDOS a mere COMMAND.COM replacement such as 4DOS, we could examine command line parsing, program loading, or batch file execution. However, as the system contains code from the master boot record on up, the whole of DOS is laid out before us. Chapters cover the structure of DOS, the boot process, file functions, disk reading and writing, file management, process management, memory management, and the command processor.

The text of the book is generally quite good, but gaps in the material are evident. In some cases, this is due to an unfortunate tendency to introduce acronyms early in the text, often without explanation and definition. (A number of these are covered later in the book.) Other missing content is only likely to be covered by examination of the source code.

DOS Internals
Geoff Chappell
Publisher: Addison-Wesley Publishing Company
1 Jacob Way, Reading, MA 01867-9984
Year: 1994
ISBN: 0-201-60835-9

Chappell admits to a rather ragged manuscript in the preparation of the book. That still shows in some of the organization and structure. The sequence of activities involved in the boot process is described in detail—but not all of that material is contained in Chapter 3, titled, "The Startup Sequence." There are also gaps in the material, even at 768 pages. Chapter 5, on "Program Execution" doesn't touch on batch files at all. Two of the four parts deal with memory management and extended memory management. The other two discuss some development tools and disk management. However, there is much here for the serious DOS programmer.

The IBM Personal Computer From the Inside Out
Murray Sargent
Publisher: Addison-Wesley Publishing Company
1 Jacob Way, Reading, MA 01867-9984
Year: 1986
ISBN: 0-201-06918-0

The title might be a bit misleading, here. You will not learn very much that will help you with hardware repairs or configuration. Nor will you learn any deep, dark, undocumented secrets. What you will get are the hardware and software fundamentals as related to BIOS/Intel/MS-DOS computers.

In spite of its age, this is still a very useful book for those who want to learn the innermost, or perhaps bottom-most workings of their PC. In spite of improvements in both Intel CPUs and MS-DOS, at heart the systems are still compatible with the earliest and oldest IBM PCs. Digital electronics have changed not at all and, in spite of new registers and opcodes, the X86 processors still contain all the workings of their predecessors. Even if you do want to take full advantage of the additional functions on the newer CPUs, you still have to learn the basics first, and this is a good place to start.

There are some obvious gaps. The introduction to assembly language would benefit enormously from a quick opcode reference chart. The coverage of digital electronics is, if not precisely elementary, at least clear and detailed. However far removed from usefulness in a PC, the construction of a simple device such as a "two-bit adder" might help give credence to the operations of computer logic.

Undocumented DOS, A Programmer's Guide to Reserved MS-DOS Functions and Data Structures
Andrew Schulman et al.
Publisher: Addison-Wesley Publishing Company
1 Jacob Way, Reading, MA 01867-9984
Year: 1994
ISBN: 0-201-63287-X

If you want the deep secrets of DOS internals, this is the book to get. For those interested in the "inside dope," this is the only game in town.

Many other authors would rest on those laurels. You want it? Come and get it. You don't like how it's done? Tough. Not so, with Schulman and company. As the Preface points out, the real purpose is to show you how DOS works. The repeated refrain is that if there is a way to do it with the documented stuff, don't take the undocumented shortcuts.

In addition, the book is carefully and clearly presented in all places. Indeed, with the added bits of history and trivia, even the nonprogrammer can find several enthralling items per chapter. I may, of course, be predisposed to it by my work in virus research, but the opening material on the investigation of Microsoft under the antitrust laws was absolutely riveting. (I said in regard to another book that I had a hard time imagining a data security text as a page-turner. Now I find myself making almost the same claim for an internals text.)

This book is definitely for programmers, and preferably those easily comfortable with both C and assembler. Unlike many useful texts, though, it won't be a drag to read.

Unauthorized Windows 95
Andrew Schulman
Publisher: IDG Books
155 Bovet Road, Suite 310, San Mateo, CA 94402
Year: 1994
ISBN: 1-56884-305-4

You will not get tips or utilities for easy Windows 95 programming. Detailed study of the book probably will, however, make you a better Windows 95 programmer. The text and associated programs examine Windows 95, with reference to earlier versions of DOS and Windows.

This book is definitely *not* for the user, and probably not even for the neophyte programmer. For the serious student of DOS and Windows internals, it is quite valuable.

System BIOS for IBM PCs, Compatibles, and EISA Computers, Second Edition
Phoenix Technologies Ltd.
Publisher: Addison-Wesley Publishing Company
1 Jacob Way, Reading, MA 01867-9984
Year: 1991
ISBN: 0-201-57760-7

On the IBM PC, the minimal amount of programming built into a computer, enough to load the rest of the operating system off the disk was referred to as the basic input/output system (BIOS). The Intel CPU at the heart of the PC was designed to work well with a system of interrupts, and so the PC BIOS was designed with these interrupts in mind. It sets up an interrupt table with a series of subroutines dealing with video (such as the setting of cursor position), disk (such as reading disk status), port (such as receiving a character), and keyboard (such as storing key data) services. It also starts a routine that loads bootstrap programming from the disk. An additional benefit of this architecture was that, within limits, certain changes could be made to the hardware. Ordinarily, such changes would require new versions of the operating system, but with the BIOS design, only the BIOS would have to be modified.

The BIOS is a complete, though limited operating system. It also forms, in a sense, the microkernel of the MS-DOS operating system, and therefore, for system level programming, it is pretty much "required reading" for MS-DOS programmers.

This reference is written by Phoenix, one of the major BIOS manufacturers. Do be aware that some minor differences may be present in other BIOS versions.

The Indispensable PC Hardware Book, Third Edition
Hans-Peter Messmer
Publisher: Addison-Wesley Publishing Company, Inc.
P.O. Box 520, 26 Prince Andrew Place
Don Mills, Ontario M3C 2T8

Year: 1997
ISBN: 0-201-40399-4

I'm not sure if this book is indispensable, but it certainly is exhaustive. The fact that it is over 1,000 pages gives you only an indication, until you see the tiny type size used in the book.

This is not a shopper's guide to the normal level of components. You get all the facts, it is true, but it is mostly at the level of pinouts and timing diagrams. This is for programmers and hardware hackers at the circuit board level and below.

An introductory chapter at a very easy and readable level leads into details of the motherboard and various component chips. Topics covered include Intel microprocessors, logical and physical memory addressing, logic gates and microprogramming, math coprocessors, derivative processors, cache, Intel "clone" processors, PC architectures and bus systems, AT architecture, extended industry standard architecture (EISA), Microchannel, peripheral component interconnect (PCI), VESA Local Bus, hardware interrupts and interrupt controller, 825× timer, direct memory access (DMA), floppy drives, hard drives, peripheral interfaces, local area network adapters, keyboards and mice, graphics adapters, and multimedia. A number of appendices bring together related information, including various listings of interrupt calls.

Given the broad scope of the book, I found some of the "missing" information to be odd. There is a listing of the machine instructions for the 80×86 processors—but only the mnemonics, without the actual opcodes or basic descriptions. The serial ports and UARTs are described thoroughly as to pinouts, but the onerous task of ensuring against address and IRQ conflicts is not discussed.

The author insists that even beginners could read this book—and he has every right to do so. If you are interested in the hardware at this level, the explanations are clear and well sequenced. For anyone curious about any of the low-level operations of the computer, this is a very thorough resource.

PC Interrupts
Ralf Brown and Jim Kyle
Publisher: Addison-Wesley Publishing Company, Inc.
1 Jacob Way, Reading, MA 01867-9984

Year: 1994
ISBN: 0-201-62485-0

Ralf Brown's MS-DOS Interrupt List is an important resource. For developers trying to avoid interrupt conflicts with other software, it is essential. For students of DOS internals, it is mandatory. For PC support staff, it is a valuable guide in determining the cause of odd behavior. In the virus research community, we are continually referring to it, and referring others to it, to determine why a virus has hooked certain interrupts, and what it's going to trigger on.

(Brown returns the compliment. He gave us Chapter 34 in the original edition, and now Chapter 59. Many names in the acknowledgments section are familiar as the more technical of the MS-DOS crowd in antiviral research.)

The Interrupt List itself is widely available on the Internet (see the reference later in this section). Because you can get the file anyway, and a more up-to-date version at that, why pay for the book? There is an additional validity to the preference for the online version in that, although there is an index to the volume, online searching is handy. However, an online reference is not always handy. You could print your own copy, but be warned, there are almost 1,000 pages of very fine print in the book.

Upgrading and Repairing PCs, Thirteenth Edition
Scott Mueller
Publisher: Macmillan Computer Publishing (MCP)
201 W. 103rd Street, Indianapolis, IN 46290
Year: 1999
ISBN: 0-7897-2542-8

There are all kinds of computer help, repair, maintenance, troubleshooting, and upgrading books on the market. A great many try to give you a quick overview of what you need to know. With the PC market expanding its options on an almost daily basis, though, generally what you need is more in the line of an encyclopedia. Your particular problem tends to be the one left out of the quick texts. This book, however, leaves very little out.

The work has a short history of the PC, the defining characteristics and components of a PC, microprocessor information (including tables and lists of detailed processor specifications, and the socket sizes and specifications), the motherboard, form factors, chipsets, interface connectors, bus sockets, the BIOS, the various types and functions of memory, the integrated drive electronics (IDE) and small computer systems interface (SCSI) interfaces, general principles of magnetic storage, specifics of hard and floppy disks, removable storage, optical drives, drive installation, display hardware, audio hardware, specifics on I/O ports, serial and parallel ports, port replacement technologies, storage interfaces, keyboards and mice, a broad range of communications hardware, local area networks, power supplies, the NVRAM (better known, if slightly inaccurately, as CMOS) battery, uninterruptible power supply (UPS) systems, portable computers, building a system, diagnostics, testing, maintenance, file systems, and data recovery. The appendices in this edition are rather curtailed. However, the CD-ROM contains full versions of the sixth, eighth, tenth, eleventh, and twelfth editions, so missing chapters, such as those on printers and software troubleshooting, can still be found.

I can say with assurance that none of the books on upgrading or repair of personal computers has had the scope of this one. This is not simply due to the size, although that certainly helps. The material is readable and clear, and there is very little fluff. Certainly some sections are not quite up to the overall standard—in particular, more recent technologies tend to have hastily assembled entries—but for the central unit itself, the book is without peer.

Inside the Windows 95 Registry
Ron Petrusha
Publisher: O'Reilly & Associates, Inc.
103 Morris Street, Suite A, Sebastopol, CA 95472
Year: 1996
ISBN: 1-56592-170-4

Petrusha's objective is to provide information for software developers creating registry enabled applications. He certainly does that, and developers should be glad. However, intermediate to advanced Windows 95 users will find this book to be worth some investigation as well. Just as DOS users were getting used to searching for and modifying .INI files, Windows 95 collects everything into the registry and changes the configuration requirements once again.

You will want to follow along in your system as you work with the book. Petrusha's explanations are designed with the developer in mind and can be terse. Nevertheless, the small systems administrator or power user will find a lot of useful information in this book.

Managing the Windows NT Registry
Paul Robichaux
Publisher: O'Reilly & Associates, Inc.
103 Morris Street, Suite A, Sebastopol, CA 95472
Year: 1998
ISBN: 1-56592-378-2

The book introduces the Registry historically, conceptually, and functionally. The history and concepts are better than the functions that are used as initial examples. At this point in the book, the actions taken by or through the Registry are rather vague. The structure of the Registry is presented very well, and there are even a number of helpful tips for values to set for specific common problems. If you know anything about the Registry, you know that there are dire warnings (and sometimes even more dire penalties) for mucking about with it. Backup and recovery are dealt with extensively and in detail, in rather stark opposition to many Registry books that simply tell you to backup. The RegEdit and RegEdt32 programs are covered with respect to the functions, strengths, and weaknesses of each.

The first of the sections dealing with Registry content is the material on system policies in Chapter 6. The explanations are very good, and better than the contents of many Windows NT security and administration texts. It can also be seen as the last chapter on tools because the policy editor is examined.

While not a complete programmer's catalogue such as the *Windows NT 4.0 Registry* by Thomas, this book contains everything that the normal user or administrator would need.

Windows NT 4.0 Registry: A Professional Reference
Steven B. Thomas
Publisher: McGraw-Hill Ryerson/Osborne
300 Water Street, Whitby, Ontario L1N 9B6
Year: 1998
ISBN: 0-07-913655-9

There are a number of other thinner books on the Registry. What sets this one apart is the inclusion of all the Registry entries you are likely to find in a basic Windows NT installation. There is also some handy Registry management software included on the CD-ROM.

The bulk of the book is in the operating system Registry entry list in part two. Chapter 9 looks at subsystem configuration in services, devices, control, and distributed, common object model (DCOM) properties. Filesystem entries are discussed in Chapter 10. Chapter 11 deals with performance monitoring functions. Network chapters are rather distributed: general services in Chapter 12, interoperability in 13, network protocols in 14, remote access in 16, routing in 17, and TCP/IP in 20. Chapter 15 covers printer entries. Security and related functions and properties are listed in Chapter 18. Chapter 19 is titled simply "Registry Problems." However, this very short chapter lists only a few, albeit important, disasters. MS Office 95 and 97 keys are listed in Chapter 21.

While a number of other books do a good job of presenting the Registry—its purpose, functions, and modification—this is the most complete in terms of the contents of the Registry itself. Given that the Registry is the descendant of the AUTOEXEC.BAT and CONFIG.SYS files, via the INIs, it is possible to argue that this book should be of interest to any intermediate NT user, and, by extension, the users of the follow-on systems of Windows 2000, Windows XP, and Windows 2003.

Windows NT Registry: A Settings Reference
Sandra Osborne
Publisher: Macmillan Computer Publishing (MCP)
201 W. 103rd Street, Indianapolis, IN 46290
Year: 1998
ISBN: 1-56205-941-6

In this case, the title is exact and the subtitle explicit. This book is about the Windows NT Registry and the settings contained therein. The material does not deal with Registry programming, which might be needed by those creating installation systems, but is of use and guidance to those involved in NT configuration, system administration, and workstation setup.

There is very little material on the Registry itself because the author assumes that those interested in buying such a book know about the existence of the system and have a rough idea of what it contains. The explanation is quick, but reasonably thorough, covering differences between the two editors, and admitting (for once) that "hive" means very little more than "file."

Part one deals with hardware settings, reviewing disk drives, display settings, mice and keyboards, sound, memory, network adapters, printers, notebooks, and other hardware issues. NT software components are covered in part two. Chapter 10 is a first-rate explanation of the NT boot process and, for any NT troubleshooter, well worth the price of the book, all by itself. Other topics include the desktop, system components, profiles, security, and the user manager settings. Part three talks about network client issues, such as dial-up networking, services, communications, and TCP/IP. An Appendix gives a complete listing of Registry keys.

Inside the Windows 95 File System
Stan Mitchell
Publisher: O'Reilly & Associates, Inc.
103 Morris Street, Suite A, Sebastopol, CA 95472
Year: 1997
ISBN: 1-56592-200-X

Data recovery people might be just a tad disappointed. This is not actually about the physical file system of sectors and tracks per se, but about file system management inside Win95. The file system that the user or application sees may be remarkably different from any physical device layout, and may, in fact, partake of a number of devices.

Chapter 1's title of "From IFSMgr to the Internet" is not exactly hype, and works on two levels. The first is an explanation of the Installable File System Manager and its provision for the management of local storage, resource access by packet requests over a network, and resources dealt with as (byte) serial data. (Hmmmm. Is Microsoft doing a UNIX on us and seeing absolutely everything as a file?) The second is a demonstration run with the author's own (and provided) MultiMon utility to learn what we can about file system activity from a session with an Internet browser.

Chapter 2 gives more sample sessions, but with strictly limited programs for more precise review of specific calls. Using MultiMon to trace the boot process, and the various ways IFSMgr is used by different types of applications, is covered in Chapter 3. File system applications programming interfaces (APIs) are detailed in Chapter

4 for Win32, KERNEL32, and Win16. Chapter 5 looks at the interrupts used by DOS.

Chapter 6 starts to look at the installable parts of the file system, the file system drivers (FSDs), and the requests that can be dispatched to them. Monitoring of file activity and more parts of MultiMon are in Chapter 7. The structure and characteristics of FSDs are detailed in Chapter 8, which also gives a sample driver that can be used to drive a monochrome monitor. Chapter 9 looks at the VFAT system, including some structural details on FAT32. Virtual memory in Win95 (and the reason its swap file is not a swap file) is examined in Chapter 10. Caching is reviewed in Chapter 11. A survey of IFSMgr services, and some of the things the documentation doesn't tell you, is included in Chapter 12. Chapter 13 talks about the network client software, with some comments on the proposed common Internet file system (CIFS). Chapter 14 looks to the future, and particularly to the relevant aspects of the Windows NT file system. Appendices provide more documentation on MultiMon, IFSMgr data structures, IFS development aids, and a Bibliography.

This is definitely a programming, rather than system support or administration manual, although some of the functions could be useful in diagnosing problems. Those who are working on Win9x applications that make extensive use of the file system will undoubtedly find much assistance in this book.

Windows NT File System Internals: A Developer's Guide
Rajeev Nagar
Publisher: O'Reilly & Associates, Inc.
103 Morris Street, Suite A, Sebastopol, CA 95472
Year: 1997
ISBN: 1-56592-249-2

This book is aimed at programmers, and they want excruciating detail. Nagar, though, provides much more than simply detail. A thorough analysis in the book makes it much more valuable than any mere list of application programming interface (API) calls. Even the first overview chapter contains sidebar notes, having little or nothing to do with the file system, that point out fascinating and possibly vital aspects to programming the system. Therefore, while the text is aimed at the developer of file systems drivers, it contains a lot of valuable information for those interested in Windows NT internals. (I was delighted to find a detailed explanation of the boot sequence for NT.)

Part one provides an overview, looking at Windows NT system components (concepts, kernel, and executive), file driver development, and structured driver development. Part two examines the NT managers: the I/O manager, virtual memory manager, and three chapters on the cache manager. Part three has three progressive chapters on the writing of file system drivers. These move through driver entry to file creation, reading, and writing, on to file information, directory control, cleanup, and file closure, and eventually to file buffer flushing, volume information, byte-range locks, opportunistic locking, device control, and file system recognizers. A final chapter covers filter drivers. Appendices deal with Windows NT system services, multiple provider router (MPR) support, kernel mode drivers, debugging support, recommended readings, and additional sources of help.

There is a great deal about Windows NT (and subsequent systems) that needs to be documented and explained. This book provides an enormous amount of valuable information for developers. More than that, however, Nagar has been able to produce clear and readable text about this vital topic, and his work will be of use to many nonprogrammers who need this level of detail and analysis.

For those who want the NASM-IDE without buying the Duntemann book, the open source assembler and an editor/development environment are available for DOS and Linux at http://uk.geocities.com/rob_anderton/. (The assembler itself can be found through a link at this site or by searching for the NASM name.) Documentation for NASM can be found at various sites, including http://www.cs.uaf.edu/~cs301/nasmdoc0.htm.

For practice in assembly language, you may find it easier to use simulators and emulators of early computers, rather than try to discover what is happening in a computer running Windows. A number of examples can be found at http://www.incolor.inetnebr.com/bill_r/computer_simulators.htm.

DEBUG is a very basic tool, but it is available with all DOS and Windows systems. To get the documentation, you do have to find a

DOS manual prior to version 4, or give the "?" command within DEBUG itself. A very detailed tutorial on DEBUG is available from Fran Golden at http://www.datainstitute.com/chap1-11.htm.

A very basic resource for assembly programming and particularly disassembly is the "x86 Interrupt List," also known as "Ralf Brown's Interrupt List" or "RBIL," and available at http://www-2.cs.cmu.edu/afs/cs.cmu.edu/user/ralf/pub/WWW/files.html.

Another important resource is the Intel architecture manual, especially part two, covering opcodes. It is available at http://developer.intel.com/design/pentiumii/manuals/243191.htm.

A very small sample program for disassembly practice is the EICAR signature test file. This is a working program that uses only printable characters. It can be created by copying the following 68 characters:

```
X5O!P%@AP[4\PZX54(P^)7CC)7}$EICAR-STANDARD-ANTIVIRUS-TEST- FILE!$H+H*
```

exactly as shown, into a file, which can then be named with an executable extension, such as .COM or .EXE. More information on the file (and various downloadable versions) can be found at EICAR on the page http://www.eicar.org/anti_virus_test_file.htm.

DEBUG has all the basic functions needed for forensic programming and analysis, but a number of other tools are easier to use for certain aspects. For example, NASM, noted previously, has a disassembler. The makers of the IDA Pro disassembler also have a free version, available at http://www.datarescue.com/freestuffuk.htm. Frhed (http://theoutcasts.net/loki/fhreddy.zip) is a free hex editor that is easier to use to view contents of files.

Collections of tools and resources related to assembly and disassembly can be found at the Fravia site (http://www.woodmann.com/fravia/tools.htm), the Assembler is not for Dummies page (http://www.softpanorama.org/Lang/assembler.shtml), Programmers Heaven (http://www.programmersheaven.com/zone5/cat25/index.htm), Protools (http://linux20368.dn.net/protools/decompilers.htm), and Foundstone (http://www.foundstone.com/knowledge/free_tools.html, which also includes tools by Robin Keir: http://keir.net/software.html). Some forensic tools are available from Sysinternals (http://www.systernals.com/index.shtml) and Winalysis (http://www.winalysis.com/).

A paper on reverse engineering malware, by Lenny Zeltser (http://zeltser.com/sans/gcih-practical/revmalw.pdf), follows the process of "black box" testing a network trojan.

For those seriously interested in getting good enough at assembly or machine language to perform forensic programming analysis of object code (and, incidentally, judging the skill level of the programmer of a piece of unknown code), an interesting work is:

Hacker's Delight
Henry S. Warren, Jr.
Publisher: Addison-Wesley Publishing Co.
P.O. Box 520, 26 Prince Andrew Place
Don Mills, Ontario M3C 2T8
Year: 2003
ISBN: 0-201-91465-4

First, this book has nothing to do with security. Experienced programmers develop toolkits of shortcut tricks. Assembly language programmers need to have more shortcuts because: a) assembly language is a detailed (and can be a tedious) process, and b) assembler gets right down to the metal, and can perform the functions that haven't been built into high level languages yet. In these days of bloatware, it may seem pointless to try to create code efficiencies. But the hackers still (sometimes) walk among us, making most efficient use of memory for code space, and squeezing every last cycle of performance out of processors. This book is for them, and anyone who wants to join them.

It is a compendium, even an encyclopedia, of such tricks. There are outlines of quick algorithms for basic operations, powers of two, bounds, counting of bits, searching words (computer "words"), transpositions and permutations, mathematical functions, and more.

This is all well and good, but isn't machine-level programming highly machine specific? Yes, to an extent. But remember, these are algorithms. The author has established them on a basic instruction set that should be common to most systems and is adaptable to situations.

Actually, as we have discussed, it does have to do with security, or at least software forensics. Not only does the security common body of knowledge have an application development domain, but this book could also be useful in forensics to gauge the skill level of a programmer.

Advanced Tools

Most of the advanced tools discussed in this book (*Software Forensics*) are available, in some form or other, online.

David Eriksson's decompiler program, Desquirr, is available at http://www.2good.com/software/desquirr/. Dcc is an older decompiler written by Cristina Cifuentes and available from http://www.it.uq.edu.au/groups/csm-old/dcc.html. You may also be interested in the University of Queensland Binary Translator (UQBT, http://www.itee.uq.edu.au/~cristina/uqbt.html) and also Boomerang, available from http://boomerang.sourceforge.net/.

In plagiarism detection, JPlag, from the University of Karlsruhe, is publicly available at http://www.jplag.de. The Berne work on finding duplicated code can be obtained at http://www.iam.unibe.ch/_scg/.

Law and Ethics

It is difficult to find good references to the law and technology. Many American authors refer only to specific U.S. laws, and ignore general legal concepts and statutes outside the United States. The law within the U.S. is in considerable flux at the moment. This lack of stability is made worse by the fact that each state is permitted to establish new criminal law, so arguments in one state may not be suitable in another.

As this book is written, probably the most useful text is:

Cyberlaw: National and International Perspectives
Roy J. Girasa
Publisher: Prentice Hall
One Lake Street, Upper Saddle River, NJ 07458
Year: 2002
ISBN: 0-13-065564-3

The back cover states that this is the "most comprehensive Internet law text for students of any discipline." The preface doesn't really contradict that statement, but then, it doesn't really specify a particular audience. The text itself, on the other hand, does not appear to be a reference, but rather a textbook for law students, and law students only. (American law students, at that.)

Part one provides an introduction and examines jurisdiction. Chapter 1 is an introduction and overview of both the technology and law. This demonstrates a number of limitations (the technology is restricted to the Internet), and, of course, the sort of bias one would expect to see in a legal text. (The definition of the Internet is taken from a "Finding of Fact" in the case that struck down the Communications Decency Act and contains a number of errors in terminology and, well, fact. The legal system is described only in terms of the various levels of U.S. courts.) A number of cases are presented regarding jurisdiction, first between U.S. states and then between states and foreign states.

Part two deals with contracts, torts, and criminal law aspects of cyberspace. The book looks at U.S. case law regarding contracts and torts, including related topics such as commercial codes like UCITA. (Many implications of the legislation are poorly expressed. There are several paragraphs describing the implied warranties under UCITA, and a brief mention of the fact that using the words "as is" voids them all.) There is also a review of international statutes dealing with commercial online transactions, and then torts, censorship, obscenity, and fraud.

Part three looks at intellectual property rights. Many of the copyright cases, all from the United States, deal with general issues not directly related to technology. Trademarks, cybersquatting (the practice of registering a domain name using a famous name or trademark, so that the owner must buy it from you), patents, and trade secrets are covered.

Part four addresses privacy and security issues, probably not having much application to software forensics.

Part five talks about antitrust, securities regulation, and relaxation. Antitrust covers Microsoft, IBM, and a number of other cases. (Analysis of the intent of programming practices possibly should have played a bigger role in the Microsoft trial.) Chapter 12's review of securities regulation cases primarily deals with fraud, and the technical links are basically irrelevant. The taxation of net businesses is also covered.

As a textbook for American law school students, this is undoubtedly useful. The cases are collected, and questions are asked to encourage students to think about various aspects of cases and related precedents that might be applicable. While U.S. structures and law predominate, there is not only acknowledgment of foreign legislation, but some detailed case examination as well. In fact, practicing lawyers would also find this volume extremely valuable, for the direction in terms of case research on precedent if nothing else. For nonlawyers, such as security professionals, the content is extremely frustrating: all questions and no answers. Still, given the extremely murky state of U.S. law in regard to the Internet and technology, this tome certainly could be worthwhile, even for those outside the United States legal system.

While it is not of practical use in dealing with specific legislation and questions of evidence, *Borders in Cyberspace* is thought-provoking and well worth a read.

Borders in Cyberspace: Information Policy and the Global Information Infrastructure
Brian Kahin and Charles Nesson
Publisher: MIT Press
55 Hayward Street, Cambridge, MA 02142-1399
Year: 1997
ISBN: 0-262-61126-0

It is rare indeed to find a collection of essays on a popular topic where each paper presents fresh insight, and all present a thorough analysis. However, Kahin and Nesson have managed to compile just such a set. Examining national and transnational issues and interests in light of the emerging global information infrastructure, each piece is informative and thought-provoking.

In part one, the essays concentrate on examining how a legal system might function on the Internet. In *The Rise of Law on the Global Network*, Johnson and Post jump completely away from the arguments about which country's legal system, all of them based on geography, should hold sway in cyberspace by suggesting that cyberspace is a separate "place," and should thus have its own laws.

Volkmer examines a number of mostly journalistic considerations regarding global versus special interest in "Universalism and Particularism." The problem of law on the 'net is revisited by Reidenberg in "Governing Networks and Rule-Making in Cyberspace," with specific emphasis on the self-regulating nature of the Internet. It has become almost a tenet of faith that increased communication brings increased democracy: faith, because hard data was unavailable. In "The Third Waves," Kedzie presents and analyzes the statistics that prove the dogma rests on a solid foundation. In "The Internet as a Source of Regulatory Arbitrage," Froomkin points out what the U.S. government has yet to learn: Cyberspace in itself has the power to enable citizens of a given jurisdiction to avoid arbitrary fiats. However, the 'net is not completely lawless, and in "Jurisdiction in Cyberspace: Intermediaries," Perritt begins to investigate a model of a court or, at least, an arbitration system.

Part two looks at issues of the conflict with geographic and political borders. Burk, in "The Market for Digital Piracy," looks at the network technologies that work against copyright law, and the various economic models that might affect this struggle. Freedom of speech and the realities of moderate censorship are examined in "A Regulatory Web: Free Speech and the GII," by Mayer-Schonberger and Foster. Gellman examines the varied outlooks on personal privacy in "Conflict and Overlap in Privacy Regulation." As has been amply demonstrated in recent years, and as Barth and Smith point out in "International Regulation of Encryption," attempts to control encryption technology are a losing battle. Governments, strapped for cash, are now trying to make money off information they have collected and formerly disseminated freely. In "International Information Policy in Conflict," Weiss and Backlund look at the various interests involved. In "Netting the Cybershark" (wonderful pun), Goldring looks at the slightly more personal topic of fraud and consumer protection in cyberspace.

Legal minds, netizens, legislators, techies, and regulators will all find something of interest in this book.

A paper on forensic linguistics at http://www.forensic-evidence.com/site/ID/linquistics.html primarily looks at the reasons why expert testimony may or may not be accepted. Another looks at proposed standards for digital evidence at http://www.fbi.gov/hq/lab/fsc/backissu/april2000/swgde.htm. (Note that these standards undergo fairly constant revision. As this book is being written, there is word of another draft to be released shortly.) There is also a list of resources for forensic linguistics at http://www.outreach.utk.edu/ljp/iafl/1997/abstracts/electronic_resources.htm.

There is a site that discusses recent developments in U.S. law at http://www.law.wayne.edu/litman/classes/cyber/newdev01s.html.

Promotional information on the Uniform Computer Information Transactions Act (UCITA) is available at UCITAOnline (http://www.ucitaonline.com/), with somewhat more substantive material at the University of Pennsylvania (http://www.law.upenn.edu/bll/ulc/ucita/ucita01.htm). In addition, a recent United States Act is said to allow copyright holders to attack sites that may be helping to break copyright. There are Web pages for the Bill itself (in PDF format at http://www.house.gov/berman/p2p.pdf) and a press release with links to the Bill and discussion of it at http://www.house.gov/berman/pr072502.htm.

The best reference in computer ethics is probably still the following:

Computer Ethics
Deborah Johnson
Publisher: Prentice Hall
113 Sylvan Avenue, Englewood Cliffs, NJ 07632
Year: 1994
ISBN: 0-13-290339-3

Unlike the famous quote about life in the state of nature being nasty, dull, brutish, and short, Johnson's examination of the state of ethics in computing is readable, interesting, discerning—and short.

The usual treatment of ethics works with proof by exhaustion, but Johnson does a complete and reasonable job. Without recourse to mounds of collected work (of dubious merit), the major points of professionalism, property rights, privacy, crime, and responsibility are addressed. Even in this brief space, ethics are studied more rigorously than in more weighty tomes. Not content with the usual reliance on relativism and utilitarianism, Johnson points out the flaws in each.

"Complete" is, I suppose, an overstatement. Although it is difficult to imagine a scenario that the book does not touch upon at some point, ultimately this book is a good primer and discussion starter. Although the definitive work in the field to date, it does not, in the final analysis, get us much closer to a computer ethic.

This work should be required reading for all computer science students. Exposure wouldn't hurt any number of professionals and executives, either.

However odd it may seem, a simplistic work that nevertheless raises most of the important questions is:

Internet and Computer Ethics for Kids
Winn Schwartau
Publisher: Inter.Pact Press
11511 Pine Street North, Seminole, FL 33772
Year: 2001
ISBN: 0-9628700-5-6

Computer ethics can be a very frustrating field. Professional organizations appear to have abandoned the area. They seem to have given up on the idea of "codes of ethics" and now prefer to write "codes of conduct." "Values education" has progressed very little in the last 30 years. All of us seem to be the disciples of Kohlberg, and assume that by sitting around discussing ethics, moral dilemmas, and scenarios, we will all somehow become moral individuals. And that's for the adults.

For kids, the task is even more important and much more difficult. Maybe it's impossible. But it is good to see that someone has at least given it a try. I don't agree with everything Winn has done, but he has produced a valuable and helpful tool. I hope that a great many people try it out, and, if it needs tuning, feed ideas back to improve it.

This volume is a tool, and must be seen as such to be valued. Schwartau has, probably wisely, not attempted to provide a full examination of ethical theories or systems. The chapters are all very short: They are introductions, not expositions. (As Blaise Pascal famously noted, it takes much longer, and much more work, to write a short piece than a long one.) The text is generally possible for the sixth grade reader, and is backed up with a short section on relevant ideas from the law, topics to think about and discuss, and resources for further study and research.

A philosophical paper on the need to change the way we look at scientific credit and copyright at http://www.firstmonday.org/issues/issue6_12/mirowski/index.html raises a number of issues related to reverse engineering. Another intriguing look at the issue of copyright in relation to concepts of privacy is found in Chapter 4 of the following book:

The Transparent Society
David Brin
Publisher: Addison-Wesley Publishing Co.
P.O. Box 520, 26 Prince Andrew Place
Don Mills, Ontario M3C 2T8
Year: 1998
ISBN: 0-201-32802-X

Chapter 4 raises an extremely interesting point in relation to copyright, patent, and other legal restrictions on intellectual property, and the fact that the information age seems to have so much trouble with it. Transparency, generally seen as the loss of privacy, initially seems to threaten destruction of the idea of copyright, but ultimately may present a unique solution to maintaining its proper function.

The book is both reasonable and provocative, and makes an interesting counterpoint to much of the current discussion of privacy and technology. Discussions of the important topics of privacy and encryption are balanced and complete, providing those near to the fields with a useful primer. In addition, Brin's more controversial points are well taken and deserve serious consideration.

Viruses and Malware

As noted in Chapter 6, forensic programming was pioneered in the field of computer viruses, and a knowledge of the structures of malware can assist us in analyzing the intent of programs.

Computer Virus Handbook
Harold Joseph Highland
Publisher: Elsevier
655 Avenue of the Americas, New York, NY 10010
Year: 1990
ISBN: 0-946395-46-2

When Dr. Highland first offered to send me a copy of this work, late in 1992, he indicated that it was outdated. In some respects this is true. Some of the precautions suggested in a few of the essays that Dr. Highland did not write tend to sound quaint. As one example, with the advantage of hindsight, Jon David's 10-page antiviral review checklist contains items of little use, and has a number of important gaps. However, for the "general," rather than "specialist" audience, this work has much to recommend it. The coverage is both broad and practical, and the information, although not quite up to date, is complete and accurate as far as it goes.

There is a concise, but encompassing overview of the viral situation by William Hugh Murray. Using epidemiology as a model, he covers the broad outline of viral functions within a computing "environment," and examines some theoretical guidelines to direct the building of policy and procedures for the prevention of viral infection. The article is broadly helpful without ever pushing the relation between computer viral and human epidemiology too far.

Chapter 3 deals with history and examples of specific viral programs. In relation to software forensics, this part is an extremely valuable resource. While other works have similar sections, the quality of this segment in Highland's tome is impressive. Mention must be made of the reports by Bill Kenny of Digital Dispatch, who provides detailed and accurate descriptions of the operations of a number of viral programs. (Chapter 4 is similar, containing three reports of viral programs from other sources.)

There is also an essay by Harry de Maio titled "Viruses—A Management Issue," and it must be considered one of the "forgotten gems" of virus literature. It debunks a number of myths and raises a number of issues seldom discussed in corporate security and virus management. Also included is a collection of essays on the theoretical aspects of computer virus research and defense.

Dated as the book may be in some respects, it is still a valuable overview for those wishing to study viral programs or the defense against them, particularly in a corporate environment. While some may find the book to be "academic" in tone, it never launches into "blue sky" speculations: All of the material here is realistic.

I may be accused of bias and egocentricity in recommending:

Viruses Revealed
Robert M. Slade, David Harley, and Urs Gattiker
Publisher: McGraw-Hill Ryerson/Osborne
300 Water Street, Whitby, Ontario L1N 9B6
Year: 2001
ISBN: 0-07-213090-3

as well as:

Robert Slade's Guide to Computer Viruses, Second Edition
Robert M. Slade
Martin Gilchrist (gilchris@sccm.stanford.edu)
Publisher: Springer-Verlag
175 Fifth Avenue, New York, NY 10010
Year: 1996
ISBN: 0-387-94663-2

All I can say is that forensic programming, its use in determining authorship and intent in malware, as well as malware components and structures is discussed in both books.

A Short Course on Computer Viruses
Fred Cohen
Publisher: Wiley
5353 Dundas Street West, 4th Floor, Toronto, Ontario M9B 6H8
Year: 1994
ISBN: 0-471-00768-4

This book is fun. I mean, it starts out with the statement, "I would like to start with a formal definition," followed by about a paragraph's worth of symbolic logic, followed by, "So much for that!" I assume that the surface joke is accessible to all. For those who know of the troubles Dr. Cohen has had over the years with those who insist on an informal translation of his work, it is doubly funny. From that beginning right through to Appendix A (a joke), the light tone is maintained throughout, and it makes for a thoroughly enjoyable read.

Besides being fun, though, the book is solid material. Possibly one could raise quibbles over certain terms or minor details, but almost nothing of substance. Dr. Cohen is, of course, the grandfather of the academic study of computer viruses. He is the source of much of the seminal thinking in regard to viruses, malware, and the defense against them.

The material in the book will be accessible to any intelligent reader, regardless of the level of computer knowledge. The most benefit, however, will be to those planning data security or antiviral policies and procedures. They will find here a thoughtful, provoking, and insightful analysis.

A Pathology of Computer Viruses
David Ferbrache
Publisher: Springer-Verlag
175 Fifth Avenue, New York, NY 10010
Year: 1992
ISBN: 0-387-19610-2

This book is a broadly based and technical compendium of research and information relevant to computer virus research on a number of platforms. For those seriously interested in the study of viral programs, this is an excellent introduction.

A series of appendices give background information on the boot sequence, record and file structure, disk structure, and other related technical details for PCs, Macs, and UNIX. As well, there are contact lists and references for further research and information.

This book is not for the home user, or even for the IT manager for a small business. The material will require some dedicated study. However, the cross-platform references and the significant security perspectives on policy and procedures will be of considerable value to the larger corporation, as well as the critical virus researcher.

While there is a great deal of superficial information about malware on the World Wide Web, detailed and technical material is rare. One of the few sources of interesting content available is the Antivirus Research division of IBM Research, at http://www.research.ibm.com/antivirus/. In relation to ethics, and the question about the possibility of a beneficial virus, I would recommend Vesselin Bontchev's paper "Are 'Good' Viruses Still a Bad Idea?" available at http://www.frisk.is/~bontchev/papers/goodvir.html.

There are a number of virus "encyclopedias" on the Web. These describe the characteristics and activity of specific viral and malicious programs, and are generally provided by vendors of antiviral scanner software. The best, in terms of accuracy of information and complete coverage of the full range of extent malware, are those from F-Secure (http://www.f-secure.com/v-descs/) and Sophos (http://www.sophos.com/virusinfo/analyses/). There are a number of others of varying accuracy, such as those provided by:

Symantec:

http://www.symantec.com/avcenter/vinfodb.html

Trend Micro:

http://www.antivirus.com/vinfo/virusencyclo/

Computer Associates:

http://www.cai.com/virusinfo/encyclopedia/

Network Associates (McAfee):

http://vil.mcafee.com/

Panda:

http://www.pandasoftware.com/library/

Kaspersky Labs:

http://www.viruslist.com/eng/viruslist.asp

Rising Antivirus:

http://www.ravantivirus.com/encyclopedia/

An independent source from a portal site provides a so-called encyclopedia at http://antivirus.about.com/compute/antivirus/library/blency.htm. It is, however, limited in information.

An unusual entry is the material provided by Trusecure. It is neither particularly accurate nor complete, but it is one of the few sources of information about viruses for the Macintosh: http://www.icsalabs.com/html/communities/antivirus/macintosh/archives/macvirus/index.html

Stylistic Analysis and Linguistic Forensics

For ideas on the cusum technique, and noncontent metrics in general, a good resource is:

Analyzing for Authorship: A Guide to the Cusum Technique
Jill M. Farringdon
Publisher: University of Wales Press
6 Gwennyth Street, Cardiff, Wales CF2 4YD
Year: 1996
ISBN: 0-7083-1324-8

Cusum (or QSUM, the two terms seem to be used interchangeably in the book) is a technique for gathering evidence about the authorship of a piece of text using numeric metrics. Instead of looking at meanings or characteristic turns of phrase, the method looks at combinations of statistical patterns in writing, patterns that the writer is probably unaware of using.

Part one is an introduction and history. There is a defense and a rough idea of the process, which would be stronger if we were presented with research indicating the likelihood of two separate authors having homogeneous or indistinguishable patterns. Included is a history of statistical stylometry studies. Details of the technique are provided, somewhat weakened by errors in the arithmetic of the examples. The bases of comparison are generally sentence length in proportion to the number of short words and words starting with vowels. This may sound strange, but an analysis of general word use in English indicates that cusum is based on syntactic structures, rather than content. As an example, the book looks at "The Back Road," suspected to be authored by D. H. Lawrence, in comparison with other works known to be by

Lawrence. The reasons for the setup chosen for this comparison are not always clear.

Part two examines a range of uses for cusum. Chapter 4 considers the statistical fingerprinting of authors even over a change of literary "voice," and also notes that an editor's style can be identified. This is extended, in Chapter 5, to the ability to identify a translator. Amazingly, consistent patterns from authors survive from childhood into adulthood, as is shown with Helen Keller's writings in Chapter 6. Chapter 7 discusses the applications of cusum to a variety of writing forms, and notes that not even the use of dialect and invented languages can hide an author's signature.

Part three looks into forensic applications. Chapter 8 lists considerations for reports to be used in court. As in the consistency over time with children, Chapter 9 demonstrates that speakers and writers of English as a second language are remarkably consistent over time, and does some analysis of the identity of confessions. Chapter 10 answers criticisms of the method. It raises good points, but has a rather confused structure. One issue raised with the cusum method is that it provides a chart to be interpreted, rather than a single measure. The text notes that statistical measures and methods are available, but that the graphics were felt to be more acceptable to users.

The book finishes off with an explanation of the method from the inventor, A. Q. Morton.

Cusum is a technique that deserves further study. Despite its flaws, the book provides valuable information. A quick introduction, excerpted from the book, is available at http://hometown.aol.com/qsums.

Software Authorship Analysis

Looking for "authorship analysis" on any Web search engine will find you a plethora of links, the vast majority of which will also have something to do with Shakespeare. As noted in Chapter 8, Biblical research and criticism provide some of the most mature techniques in this area. A. Q. Morton, inventor of cusum, started his work as a Biblical scholar. An overall discussion of authorship analysis tools and techniques can be found at http://linguistlist.org/issues/5/5-1067.html. The paper on short substrings in document discrimination is at http://www.qucis.queensu.ca/achallc97/papers/p025.html.

A paper on “Authorship Analysis: Identifying the Author of a Program,” by Ivan Krsul and Gene Spafford, is available at http://citeseer.nj.nec.com/krsul96authorship.html, among other places. There are a number of versions of, and locations for this work, but using CiteSeer gets you links to essays related to the same topic. Other papers, such as the report by Gene Spafford and Stephen Webber, can be found at the CERIAS Library at http://www.cerias.purdue.edu/coast/coast-library.html.

Index

Note: Boldface numbers indicate illustrations.

About the Author

Robert M. Slade has been a security consultant since 1987, working for some of the best-known *Fortune* 500 companies and the government of Canada. The author of *Robert Slade's Guide to Computer Viruses*, and co-author of *Viruses Revealed*, he also teaches. He has prepared curricula and taught courses for Simon Fraser University, MacDonald Dettwiler and Associates, Ltd., and the University of Phoenix, among others. He is a CISSP (Certified Information Systems Security Practitioner) trainer and a specialist in malware.